FM 3-60 (FM 6-20-10)

The Targeting Process

NOVEMBER 2010

DISTRIBUTION RESTRICTION: Approved for public release; distribution is unlimited.

Headquarters, Department of the Army

*FM 3-60 (FM 6-20-10)

Field Manual
No. 3-60

Headquarters,
Department of the Army
Washington, DC, 26 November 2010

The Targeting Process

Contents

		Page
	PREFACE	iv
	INTRODUCTION	v
Chapter 1	TARGETING PRINCIPLES AND PHILOSOPHY	1-1
	Doctrinal Basis	1-1
	Targeting Principles	1-2
	Targeting Guidance	1-2
	Targeting Categories	1-4
	Targeting Methodology	1-6
	Targeting and Military DecisionMaking Process	1-7
	Service Components Targeting Methodology	1-12
	Air-Ground Operations Relationship	1-12
	Targeting Personnel and Responsibilites	1-12
Chapter 2	TARGETING METHODOLOGY—LETHAL AND NONLETHAL	2-1
	General	2-1
	Decide	2-2
	Detect	2-10
	Deliver	2-15
	Assess	2-19
Chapter 3	CORPS AND DIVISION TARGETING	3-1
	Fires Cell	3-1
	Requirements for Successful Targeting	3-5
	D3A Methodology	3-6
	Corps and Division Synchronization	3-8
	Air-Ground Operations at Corps Level	3-8
	Targeting Responsibilities	3-10
Chapter 4	BRIGADE COMBAT TEAM AND BATTALION TASK FORCE TARGETING	4-1
	Functions	4-1
	Planning Considerations	4-2
	Targeting Working Group Battle Rhythm	4-3

Distribution Restriction: Approved for public release; distribution is unlimited.

*This publication supersedes FM 6-20-10, 8 May 1996.

Contents

  BCT Targeting Working Group Membership 4-4
  BCT Targeting Board Membership 4-4
  BCT Targeting Responsibilities 4-5
  BCT Fires Cell 4-11
  Preparing and Conducting Targeting Working Group 4-23
  Synchronization 4-28

Appendix A   FIND, FIX, TRACK, TARGET, ENGAGE, AND ASSESS A-1
Appendix B   FIND, FIX, FINISH, EXPLOIT, ANALYZE, AND DISEMINATE B-1
Appendix C   NATIONAL INTELLIGENCE ORGANIZATIONS SUPPORT TO TARGETING C-1
Appendix D   EXAMPLE FORMATS AND TARGET REPORTS D-1
Appendix E   TARGETING CHECKLIST E-1
Appendix F   SAMPLE TARGETING TEAM STANDARD OPERATING PROCEDURES F-1
Appendix G   COMMON DATUM G-1
Appendix H   TARGET NUMBERING H-1
   GLOSSARY Glossary-1
   REFERENCES References-1
   INDEX Index-1

Figures

Figure 1-1. Targeting categories 1-6
Figure 1-2. Targeting methodology 1-7
Figure 1-3. D3A methodology and the MDMP 1-8
Figure 2-1. D3A methodology cycle 2-1
Figure 2-2. Targeting products (example) 2-3
Figure 2-3. High-payoff target list (example) 2-6
Figure 2-4. Target selection standards matrix (example) 2-8
Figure 2-5. Attack guidance matrix (example) 2-9
Figure 2-6. Target report 2-13
Figure 2-7. Target validation considerations 2-15
Figure 2-8. Assessment levels and measures 2-19
Figure 2-9. D3A methodology 2-23
Figure 3-1. Fires cell 3-2
Figure 3-2. FIB fires cell and elements (example) 3-17
Figure 4-1. Brigade battle rhythm (example) 4-3
Figure A-1. Find step determinations and follow on actions A-2
Figure B-1. High-value individual targeting process B-2
Figure D-1. High-payoff target list (sample) D-2
Figure D-2. Target selection standards matrix (sample) D-2

Figure D-3. Attack guidance matrix (sample) D-3
Figure D-4. Combined HPTL TSS AGM (sample) D-6
Figure D-5. Target selection standards worksheet (sample) D-7
Figure D-6. Targeting synchronization matrix (sample) D-8
Figure D-7. Combined lethal/nonlethal targeting synchronization matrix (sample) D-9
Figure D-8. Alternate targeting synchronization matrix format (sample) D-10
Figure D-9. Delivery standards matrix (sample) D-11
Figure D-10. Baseball card (front side) (sample) D-14
Figure D-11. Baseball card (backside) (sample) D-15
Figure D-12. Picture of HVI residence (sample) D-15
Figure D-13. HVI link diagram (sample) D-16
Figure D-14. HVI reports (sample) D-16
Figure D-15. HVI summary/assessment (sample) D-17
Figure F-1. Targeting working group (example) F-1
Figure F-2. Targeting working group agenda (example) F-2
Figure F-3. Meeting times interface between corps and division (example) F-3
Figure G-1. Map margin datum (example) G-2

Tables

Table 1-1. Crosswalk of operations process, joint targeting cycle, D3A, and MDMP 1-11
Table 4-1. Targeting working group agenda (example) 4-25
Table 4-2. BCT fire support execution matrix (example) 4-28
Table 4-3. Task force fire support execution matrix (example) 4-29
Table D-1. HPTL-AGM option 1 (sample) D-4
Table D-2. HPTL-AGM option 2 (sample) D-4
Table D-3. Target report (sample) D-12
Table D-4. Air tasking order (example) D-13
Table H-1. Assignment of first letter (example) H-1
Table H-2. Assignment of letters (example) H-2
Table H-3. Assignment of blocks of numbers (example) H-2
Table H-4. Additional assignment of blocks of numbers (example) H-2

Preface

PURPOSE

This publication (Field Manual [FM] 3-60, *The Targeting Process*) describes the targeting process used by the United States Army. The FM 3-60 is descriptive and not prescriptive in nature. This manual has applicability in any theater of operations. The manual offers considerations for commanders and staffers in preparing for challenges with targeting, yet it is flexible enough to adapt to dynamic situation. FM 3-60 replaces FM 6-20-10, *Tactics, Techniques, and Procedures for the Targeting Process*.

The development and research of FM 3-60 parallels similar ongoing efforts by other Army proponents to develop their own supporting branch doctrine and tactics, techniques, and procedures for the division, support brigades, brigade combat teams, and subordinate elements.

ADMINISTRATIVE INFORMATION

Unless this publication states otherwise, masculine nouns and pronouns do not refer exclusively to men.

This publication applies to the Active Army, the Army National Guard/Army National Guard of the United States, and the United States Army Reserve unless otherwise stated.

The United States Army Training and Doctrine Command is the proponent for this publication. The preparing agency is the United States Army Fires Center of Excellence and Fort Sill. Users are invited to send written comments and recommendations on a DA Form 2028 (Recommended Changes to Publications and Blank Forms) directly to Directorate of Training and Doctrine, 700 McNair Avenue, Suite 128 ATTN: ATSF-DD, Fort Sill, OK 73503; by e-mail to atsfddd@conus.army.mil; or submit an electronic DA Form 2028.

Introduction

SCOPE

Field Manual (FM) 3-60, *The Targeting Process* consists of five chapters and eight appendices to describe the Army's targeting process. Each chapter and appendix addresses how the decide, detect, deliver, and assess (D3A) methodology enhances the targeting process. The D3A is a methodology which optimizes the integration and synchronization of maneuver, fire support, and intelligence from task force to corps level operations. The D3A is described without tying it to specific hardware that will eventually become dated. The Army's targeting process consists of time tested techniques organized in a systematic framework.

The FM 3-60 addresses how D3A methodology interfaces with the joint targeting cycle, military decisionmaking process (MDMP), and operations process. The joint targeting fundamental principles and doctrinal guidance are also presented in this publication.

Successful targeting requires that the leadership team and their staff possess an understanding of the functions associated with the targeting process. The FM 3-60 builds on the collective knowledge, experience gained through recent operations, and numerous exercises. The manual is rooted in time tested principles and fundamentals, while accommodating force design, new technologies, and diverse threats to national security.

The targeting process is challenging. The challenge includes locating, identifying, classifying, tracking, and attacking targets and assessing battle damage with limited assets and weapon systems, which makes this process complicated. The process becomes even more difficult with long range and fast moving targets. It is even more complex at division and higher echelons with more decisionmakers, acquisitions, surveillance assets, and weapon systems. This challenge is particularly true when joint and combined assets are included. The competition for assets is intense. Many intelligence systems are capable of situation development, target acquisition, and battle damage assessment (BDA), but may not be able to do all at the same time. Detailed guidance, thorough planning, and disciplined execution prevent unnecessary redundancy and make the most of available combat power.

Chapter 1 begins with the basics and introduction to targeting.

Chapter 2 describes the Army's targeting process in detail.

Chapter 3 addresses targeting at the corps and division level.

Chapter 4 addresses targeting at the brigade combat team and battalion level.

Appendix A describes the joint dynamic targeting process.

Appendix B describes a method for targeting high-value individuals (HVI).

Appendix C describes the targeting support available from national agencies.

Appendix D provides examples of targeting products applicable to the operations process.

Appendix E provides a targeting checklist.

Appendix F provides a targeting working group standard/standing operating procedure.

Appendix G provides information on the common datum.

Appendix H explains the target numbering system for targeting.

This page intentionally left blank.

Chapter 1
Targeting Principles and Philosophy

According to joint publication (JP) 3-60, a *target* is an entity or object considered for possible engagement or other action. Targets also include the wide array of mobile and stationary forces, equipment, capabilities, and functions that an enemy commander can use to conduct operations. *Targeting* is the process of selecting and prioritizing targets and matching the appropriate response to them, considering operational requirements and capabilities (JP 3-0). The emphasis of targeting is on identifying resources (targets) the enemy can least afford to lose or that provide him with the greatest advantage, then further identifying the subset of those targets which must be acquired and attacked to achieve friendly success. Denying these resources to the enemy makes him vulnerable to friendly battle plans. These resources constitute critical enemy vulnerabilities. Successful targeting enables the commander to synchronize intelligence, maneuver, fire support systems, nonlethal systems, and special operations forces by attacking the right target with the best system at the right time. Targeting is a complex and multidiscipline effort that requires coordinated interaction among many command functions. These command functions in collaboration are referred to as the targeting working group and include, but are not limited to, the fires, intelligence, current operations, future operations, and plans cells. Representatives from these cells are essential to a comprehensive targeting process. Other members of the staff may help them in the planning and execution phases of targeting. Close coordination among all cells is crucial for a successful targeting effort. Sensors and collection capabilities under the control of external agencies must be closely coordinated for efficient and quick reporting of fleeting or dangerous targets. In addition, the appropriate means and munitions must attack the vulnerabilities of different types of targets.

DOCTRINAL BASIS

1-1. The Army will not operate alone in the uncertain, ambiguous security environment described in JP 3-0, JP 3-60, and field manual (FM) 3-0. Operations involving Army forces will frequently participate in joint operations. The overarching operational level considerations are for the joint force commander (JFC) to synchronize the action of air, land, sea, space, and special operations forces to achieve strategic and operational objectives through integrated, joint campaign and major operations. The JFC seeks to win decisively and quickly and with minimum casualties and minimal collateral damage. The application of scalable fires is essential in defeating the enemy's ability and will to fight. The JFC uses a variety of means to divert, limit, disrupt, delay, damage, or destroy the enemy's air, surface, and subsurface military potential throughout the joint operational area. The specific criteria of the above terms must be established by the commander and well understood by targeting working group members. Conflicts will be dominated by high technology equipment and weapons and fought over extended distances by highly integrated joint and combined task forces. The characteristics of the future battlefield will challenge the joint force and Service component commanders' ability to efficiently and effectively employ limited numbers of sophisticated acquisition and weapon systems against a diverse target array with efficiently and effectively.

1-2. Targeting is a critical process of the fires warfighting function. The *fires warfighting function* is the related tasks and systems that provide collective and coordinated use of Army indirect fires and joint fires through the targeting process (FM 3-0) It includes tasks associated with integrating and synchronizing the effects of the types of fires with the effects of other warfighting functions. Commanders integrate these tasks into the concept of operations during planning and adjust them based on the targeting guidance. Fires normally contribute to the overall effect of maneuver, but commanders may use them separately for decisive and shaping operations. The fires warfighting function includes the following tasks:

- Decide surface targets.
- Detect and locate surface targets.
- Provide fire support
- Assess effectiveness.
- Integrate and synchronize cyber-electromagnetic activities .

1-3. These tasks are integrated into the operational level during planning and adjusted based on the targeting guidance.

TARGETING PRINCIPLES

1-4. The enemy presents a large number of targets, frequently more than can be serviced with available intelligence, acquisition, and attack assets. The targeting process weighs the benefits and the cost of attacking various targets in order to determine which targets, if attacked, are most likely to contribute to achieving the desired end state. Adhering to four targeting principles should increase the probability of creating desired effects while diminishing undesired or adverse collateral effects. These principles are—

- The targeting process is focused on achieving the commander's objectives. It is the function of targeting to achieve efficiently those objectives within the parameters set at the operational level, directed limitations, the rules of engagement, or rules for the use of force, the law of war, and other guidance given by the commander. Every target nominated must contribute to attaining the commander's objectives.
- Targeting is concerned with the creation of specific desired effects through lethal and nonlethal actions. Target analysis considers all possible means to achieve desired effects, drawing from all available capabilities. The art of targeting seeks to achieve desired effects with the least risk and expenditure of time and resources.
- Targeting is a command function that requires the participation of many disciplines. This entails participation from all elements of the unit staff, special staff, special augmentees, other agencies, organizations, and multinational partners. Many of the participants may directly aid the targeting effort while working at locations vast distances from the unit. Even company level targeting elements frequently have access to intelligence and analysis generated by national elements.
- Targeting seeks to achieve effects through lethal and nonlethal actions in a systematic manner. A targeting methodology is a rational and iterative process that methodically analyzes, prioritizes, and assigns assets against targets systematically to achieve those effects that will contribute to achieving the commander's objectives. If the desired effects are not achieved, targets are recycled through the process.

TARGETING GUIDANCE

1-5. The commander's targeting guidance must be articulated simply yet authoritatively. The guidance must be easily understood across the combined and joint environment of operational areas. Targeting guidance must focus on essential enemy capabilities and functions that could interfere with the achievement of friendly objectives, such as, the ability to exercise control of forward units, the ability to mass fire support, or (in stability operations) the ability to manufacture explosive devices. These are high value targets needed by the enemy to accomplish his own mission and to keep friendly forces from achieving theirs.

1-6. The commander's targeting guidance describes the desired effects to be generated by scalable fires, physical attack, and cyber/electromagnetic activities against enemy and adversary operations. The ability to execute the listed elements in different locations at the same time producing the following desired effects— deceive, degrade, delay, deny, destroy, disrupt, divert, exploit, interdict, neutralize, and suppress. Terms such as delay, disrupt, divert, and destroy have long been used to describe the effects of artillery fire on enemy capabilities; as have terms such as destroy, disrupt, degrade, deny, deceive, and exploit to describe the effects of information operations (FM 3-13). The terms are not mutually exclusive. Action associated with one desired effect may also support other desired effects. For example, delay can result from disrupting, diverting, or destroying enemy capabilities or targets.

1-7. Various scalable fires can be employed by a commander to achieve the following effects on an enemy or adversary—

- **Deceive.** To *deceive* is to cause a person to believe what is not true (FM 3-13). Military deception seeks to mislead adversary decisionmakers by manipulating their understanding of reality. Decisionmakers can be deceived because they operate in an uncertain environment. Uncertainties about the situation and the inability to predict outcomes accurately require adversaries to take risks that can expose them to the effects of friendly fires.
- **Degrade.** To *degrade* is to use nonlethal or temporary means to reduce the effectiveness or efficiency of adversary command and control systems and information collection efforts or means (FM 3-13).
- **Delay.** To *delay* is to slow the time of arrival of enemy forces or capabilities or alter the ability of the enemy or adversary to project forces or capabilities. When enemy forces are delayed friendly forces gain time (JP 3-03). For delay to have a major impact, the enemy must face urgent movement requirements or the delay must enhance the effect(s) of friendly operations. When delayed enemy forces mass behind a damaged route segment a more concentrated set of targets and a longer period of exposure to friendly fires results. (JP 3-03).
- **Deny.** To *deny* is to withhold information about Army force capabilities and intentions that adversaries need for effective and timely decisionmaking (FM 3-13). To deny is also to hinder or prevent the enemy from using terrain, space, personnel, supplies, or facilities. Destruction of enemy equipment is an effective means of denying his use of the electromagnetic spectrum; however, the duration of denial will depend on the enemy's ability to reconstitute. The electronic warfare (EW) representative must consider unintended consequences of EW operations. Friendly electronic attack could potentially deny essential services to a local populace, which in turn could result in loss of life and/or political ramifications.
- **Destroy.** In the context of defeat mechanisms, to *destroy* is to apply lethal combat power on an enemy capability so that it can no longer perform any function and cannot be restored to a usable condition without being entirely rebuilt (FM 3-0). A building is destroyed when all vertical supports and spanning members are damaged to such an extent that nothing is salvageable. In the case of bridges, all spans must have dropped and all piers must require replacement. The amount of damage needed to render a unit combat ineffective depends on the unit's type, discipline, and morale. Any such percentages must be specified by the supported unit commander. Area fire weapons require considerable ammunition and time to destroy armored or dug in targets as direct hits with high explosive shells are generally required. Precision guided munitions are a better means of destroying such targets. When used in the EW context, destruction is the elimination of targeted adversary's systems. Sensors and command nodes are lucrative targets because their destruction strongly influences the enemy's perceptions and ability to coordinate actions. EW support supports destruction by providing target location and/or information. Adversary systems that use the electromagnetic spectrum can be destroyed by a variety of weapons and techniques, ranging from conventional munitions and directed energy weapons to network attacks.
- **Disrupt.** To *disrupt* is to interrupt or impede enemy or adversary capabilities or systems, upsetting the flow of information, operational tempo, effective interaction, or cohesion of the enemy force or those systems (JP 3-03). *Disrupt* is a tactical mission task in which a commander integrates direct and indirect fires, terrain, and obstacles to upset an enemy's formation or tempo, interrupt his timetable, or cause his forces to commit prematurely or attack in piecemeal fashion. In information operations, *disrupt* is breaking and interrupting the flow of information between selected command and control nodes (FM 3-13). Any of these may in turn cause enemy forces to commit prematurely or attack in a piecemeal fashion. Attacking command and control nodes may force the enemy to use less capable, less secure backup communications systems that can be more easily exploited by friendly force. Attacking enemy lines of communications may force the enemy to use less capable transportation modes to communicate with and sustain their

Chapter 1

forces. Uncertainty as to whether or not forces, materiel, or supplies will arrive can directly affect enemy commanders, their staffs, and forces.
- **Divert.** To *divert* is to force the enemy or adversary to change course or direction. Diversion causes enemy forces to consume resources or capabilities critical to enemy operations in a way that is advantageous to friendly operations. Diversions draw the attention of enemy forces away from critical friendly operations and prevent enemy forces and their support resources from being employed for their intended purpose. Diversions can also cause more circuitous routing along lines of communications, resulting in delays for enemy forces (JP 3-03).
- **Exploit.** To *exploit* is in information operations, to gain access to adversary command and control systems to collect information or to plant false or misleading information (FM 3-13).
- **Interdict.** To *interdict* is to divert, disrupt, delay, or destroy the enemy's military surface capability before it can be used effectively against friendly forces, or to otherwise achieve objectives (JP 3-03).
- **Neutralize.** To *neutralize* is to render enemy personnel or material incapable of interfering with a particular operation (FM 1-02). The unit is effective again when the casualties are replaced and/or damage is repaired. Any such percentages must be specified by the supported unit commander. Neutralization fires are delivered against targets located by accurate map inspection, indirect fire adjustment, or a target acquisition device. The assets required to neutralize a target vary according to the type and size of the target and the weapon ammunition combination.
- **Suppress.** To *suppress* is to temporarily degrade the performance of a force or weapons system below the level needed to accomplish the mission (FM 1-02). Firing high explosive rounds with variable time fuzes reduces the combat effectiveness of personnel and armored targets by creating apprehension and surprise and by causing tracked vehicles to button up. Obscuration is used to blind or confuse. Fires used to suppress are useful against likely, suspected, or inaccurately located enemy units where time is essential. They can be provided by small delivery units or means and require little ammunition. Suppression fires such as a smoke screen continue long enough to degrade enemy performance.

1-8. The commander can also provide restrictions as part of his targeting guidance. Targeting restrictions fall into two categories. The two categories are no-strike list and restricted target list. The no-strike list consists of objects or entities protected by the following—
- Law of armed conflict.
- International laws.
- Rules of engagement.
- Other considerations.

The no-strike list is developed independently of and in parallel to the candidate target list.

1-9. A restricted target list is a valid target with specific actions. Listed below are some examples of specific actions—
- Limit collateral damage.
- Preserve select ammo for final protective fires.
- Do not strike during daytime.
- Strike only with a certain weapon.
- Proximity to no-strike facilities.

Note. See JP 2-0 and JP 3-60 for additional information on legal considerations and targeting restrictions.

TARGETING CATEGORIES

1-10. There are two targeting categories— deliberate and dynamic.

DELIBERATE TARGETING

1-11. Deliberate targeting prosecutes planned targets. These targets are known to exist in an operational area and have actions scheduled against them. Examples range from targets on target lists in the applicable plan or order, targets detected in sufficient time to be placed in the joint air tasking cycle, mission type orders, or fire support plans.

1-12. There are two types of planned targets: scheduled and on-call–

- Scheduled targets exist in the operational environment and are located in sufficient time or prosecuted at a specific, planned time.
- On-call targets have actions planned, but not for a specific delivery time. The commander expects to locate these targets in sufficient time to execute planned actions. These targets are unique in that actions are planned against them using deliberate targeting, but execution will normally be conducted using dynamic targeting such as close air support missions and time-sensitive targets (TST).

DYNAMIC TARGETING

1-13. Dynamic targeting prosecutes targets of opportunity and changes to planned targets or objectives. Targets of opportunity are targets identified too late, or not selected for action in time, to be included in deliberate targeting. Targets prosecuted as part of dynamic targeting are previously unanticipated, unplanned, or newly detected. If the target is not critical or time-sensitive enough to warrant prosecution during the current execution period, the target may be developed for prosecution during a later execution period. Analysis of the target may also determine that no action is needed.

1-14. There are two types of targets of opportunity: unplanned and unanticipated–

- Unplanned targets are known to exist in the operational environment, but no action has been planned against them. The target may not have been detected or located in sufficient time to meet planning deadlines. Alternatively, the target may have been located, but not previously considered of sufficient importance to engage.
- Unanticipated targets are unknown or not expected to exist in the operational environment.

TIME-SENSITIVE TARGETS

1-15. A TST is a JFC designated target requiring immediate response because it is a highly lucrative, fleeting target of opportunity or it poses (or will soon pose) a danger to friendly forces (JP 3-60). TST is a JFC designated target or target type of such high importance to the accomplishment of the JFC mission and objectives or one that presents such a significant strategic or operational threat to friendly forces or allies, that the JFC dedicates intelligence collection and attack assets or is willing to divert assets away from other targets in order to find, fix, track, target, engage, and assess (F2T2EA) it/them.

1-16. TST comprises a very small or limited number of targets due to the required investment of assets and potential disruption of planned execution, and are only those targets designated by the JFC and identified as such in the JFC guidance and intent. TSTs are normally executed dynamically; however, to be successful, they require considerable deliberate planning and preparation within the joint targeting cycle.

1-17. Service component commanders may designate high-priority targets that present significant risks to or opportunities for component forces and/or missions. These are generally targets that the Service component commander(s) have nominated to the JFC TST list, but were not approved as TSTs. This class of targets may require rapid processing and cross component coordination, even though they did not qualify for inclusion on the JFC TST list. The JFC and Service component commanders should clearly designate these targets prior to execution of military operations. Such targets will generally be prosecuted using dynamic targeting. These targets should receive the highest priority possible, just below targets on the JFC TST list.

Chapter 1

SENSITIVE TARGETS

1-18. Certain targets require special care or caution in treatment because failure to attack them or to attack them improperly can lead to major adverse consequences. Example includes leadership targets (high-value individuals [HVI]) that must be handled sensitively due to potential political repercussions; targets located in areas with a high risk of collateral damage; and weapons of mass destruction facilities, where an improper attack can lead to major long-term environmental damage. Such targets are often characterized as "sensitive" in one respect or another, without having the intrinsic characteristics, by definition, of a sensitive target. Nonetheless, the manner in which they are attacked is sensitive and may require coordination with and approval from the JFC or higher authorities. In most cases, it is best to establish criteria for engaging such targets in as much detail as possible during planning, before combat commences. (See figure 1-1.)

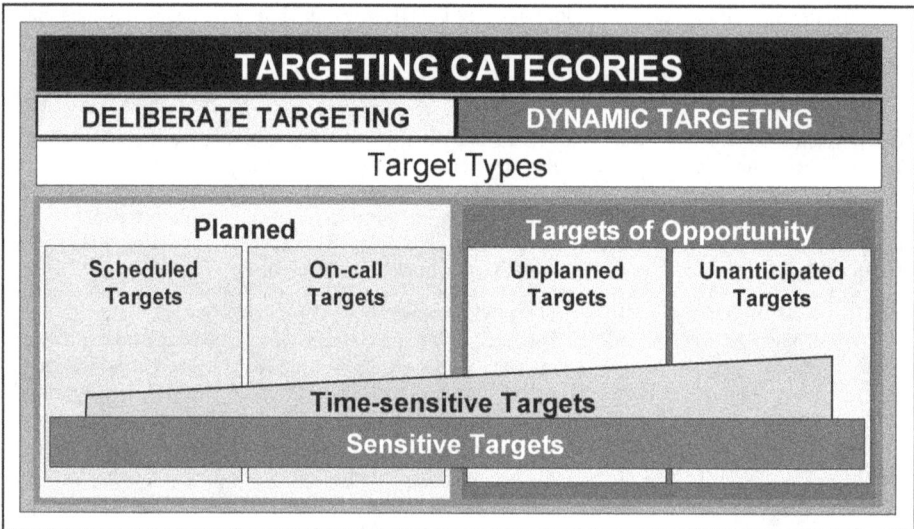

Figure 1-1. Targeting categories

TARGETING METHODOLOGY

1-19. Targeting process and the decide, detect, deliver, and assess (D3A) methodology is time tested and performed by the commander's staff in planning and executing of targets. The methodology has four functions. Details of each function are presented in Chapter 2. This methodology organizes the efforts of the commander and staff to accomplish key targeting requirements. The targeting process supports the commander's decisions. It helps the targeting working group decide which targets must be acquired and attacked. It helps in the decision of which attack option to use to engage the targets. Options can be lethal or nonlethal and/or organic or supporting at all levels through the range of operations as listed—maneuver, electronic attack, psychological, attack aircrafts, surface-to-surface fires, air to surface, or a combination of these operations. In addition, the process helps in the decision of who will engage the target at the prescribed time. It also helps targeting working groups determine requirements for combat assessment to assess targeting and attack effectiveness. (See figure 1-2.)

Targeting Principles and Philosophy

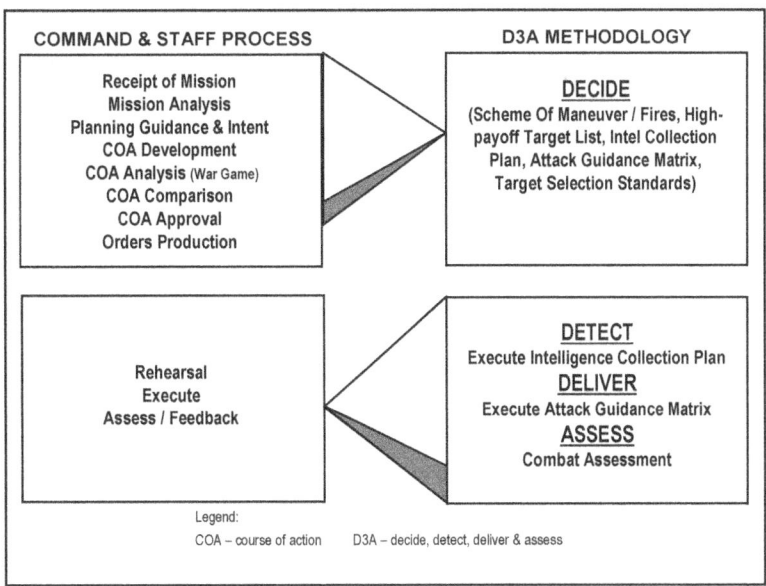

Figure 1-2. Targeting methodology

TARGETING AND MILITARY DECISIONMAKING PROCESS

1-20. The D3A methodology is an integral part of the military decisionmaking process (MDMP) from receipt of the mission through operation order (OPORD) execution and assessment. Like MDMP, targeting is a leadership driven process. Targeting frequently begins simultaneously with receipt of mission, and may even begin based on a warning order. As the MDMP is conducted, targeting becomes more focused based on the commander's guidance and intent. The composite risk management process is an integral tool in the MDMP and is compatible process that aligns with MDMP. The S-3 in units without a protection cell and the safety officer integrates the composite risk management into the MDMP. See FM 5-19 for a more detailed illustration of the first four steps of composite risk management conducted in the MDMP.

MDMP

1-21. The commander is responsible for mission analysis but may have his staff conduct a detailed mission analysis for his approval. The mission analysis considers intelligence preparation of the battlefield (IPB), environmental considerations, enemy situation, and potential enemy course of action (COA).

Note. The joint targeting cycle uses the term joint intelligence preparation of the operational environment (JIPOE).

1-22. The commander provides his initial planning guidance and intent for further COA developments. The initial guidance and intent is given in a warning order. A warning order is sent to subordinate units to allow them to begin planning, providing them as much lead time as possible.

1-23. The plans cell develops potential friendly COAs based on facts and assumptions identified during IPB and mission analysis. These developed friendly COAs are usually checked by the commander or chief of staff to ensure they comply with the commander's initial guidance and intent and meet considerations for COA development. The intelligence staff develops as many possible enemy COAs as time allows.

Chapter 1

1-24. The rules are developed by the rules of engagement cell under the supervision of the operations offices and assisted by staff judge advocate, based on commander's guidance, during the planning phase of the operations process.

1-25. Once approved for further development, a friendly COA is war gamed against the most likely and/or most threatening enemy COA to determine their suitability, acceptability, and feasibility. These results are normally briefed to the commander in a decision briefing. Following a decision by the commander, adjustments are made, if necessary, to the selected COA and orders preparation begins. A warning order is with as much information as possible to expedite their planning.

1-26. The OPORD is completed by the staff and approved by the commander and then issued to subordinates. Subordinate units continue their planning process and modifying supporting plans as necessary. Rehearsals should be conducted before execution. The order is executed, and the commander and staff assess activities and results. The assessment provides them with feedback for modifying current plans or identifying new missions.

1-27. Figure 1-3 illustrates the relationship between the D3A methodology and the MDMP along with products generated during the targeting process.

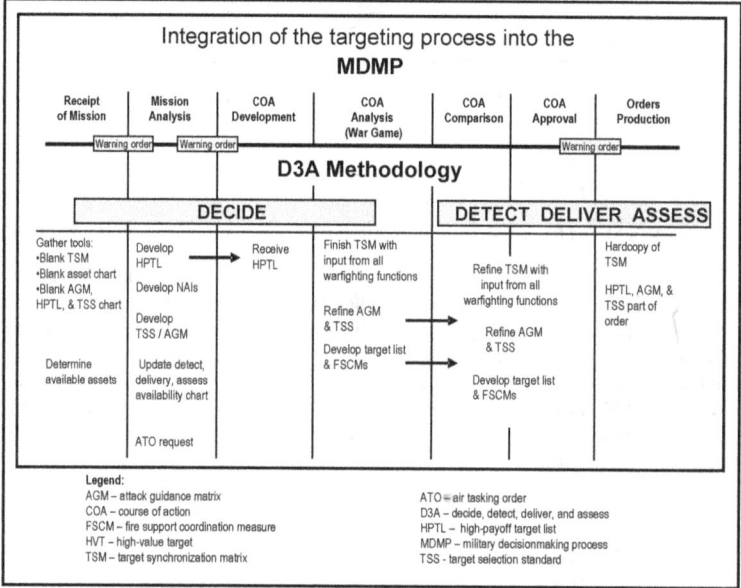

Figure 1-3. D3A methodology and the MDMP

D3A IN MDMP

1-28. The decide function coincides with the MDMP from the mission analysis through the issuing of the approved plan or order. The detect function starts with the commanders approval of the plan or order and is accomplished during execution of the plan or order. Once detected, targets are attacked and assessed as required targeting working groups are used as a vehicle to focus the targeting process within specified time.

1-29. D3A methodology functions occur simultaneously and sequentially during the operations process. While an individual target may progress through each step of the process sequentially, there are normally multiple targets in each step of the process. As decisions are made in planning future operations, the current operations staff conducts the detection, execution, and planning of targets based on prior decisions.

Decide

1-30. The decide function is the most important and requires close interaction between the commander and the intelligence, plans, operations, fires cells, and staff judge advocate. The staff officers must clearly understand the following—
- Unit mission.
- Commander's intent (scheme of maneuver and scheme of fires).
- Commander's planning guidance.
- Rules of engagement.

1-31. With this information, the staff officers can prepare their respective running estimates. From the standpoint of targeting, the fire support, intelligence, and operations estimates are interrelated and closely coordinated among each cell. Key staff products include target value analysis, and the intelligence estimate from the intelligence and targeting the intelligence officers. War gaming allows the chief of fires/fire support officer (FSO) to develop the decide function products. The decide function gives a clear picture of the priorities that apply to the following—
- Tasking of target acquisition assets.
- Information processing.
- Selection of an attack means and measures of effectiveness.
- Requirement for combat assessment.
- Target sets.

1-32. The resulting OPORD addresses key points of the decision support template. The order contains commander's critical information requirements to include the following—
- Priority intelligence requirements (PIR).
- Information requirements.
- Intelligence, surveillance, and reconnaissance (ISR) plan.
- Target acquisition tasks.
- High-payoff target list (HPTL).
- Attack guidance matrix (AGM).
- Target selection standards (TSS).
- Any requirements for assessment.

Detect

1-33. The detect targeting function is conducted during the execution of the OPORD. During detection, the assistant chief of staff, intelligence (G-2)/intelligence staff officer (S-2), and assistant chief of staff, operations (G-3)/operations staff officer (S-3) supervise the execution of the ISR plan. Target acquisition assets gather information and report their findings back to their controlling headquarters, which in turn pass pertinent information to the tasking agency. Some collection assets provide actual targets, while other assets must have their information processed to produce valid targets. Not all of the information reported would benefit the targeting effort, but it may be valuable to the development of the overall situation. The target priorities developed in the decide function are used to expedite the processing of targets. Situations arise where the attack, upon location and identification, of a target is either impossible (for example out of range) or undesirable (outside of but moving toward an advantageous location for the attack). Critical targets that we cannot or choose not to attack in accordance with the attack guidance must be tracked to ensure they are not lost. Tracking suspected targets expedites execution of the attack guidance. Tracking suspected targets keeps them in view while they are validated. Planners and executers must keep in mind that assets used for target tracking may be unavailable for target acquisition. As targets are developed, appropriate weapon systems are tasked in accordance with the attack guidance and location requirements of the system.

Deliver

1-34. The deliver function main objective is to attack targets in accordance with the attack guidance provided. The tactical solution (the selection of weapon system or a combination of weapons system) leads to a technical solution for the selected weapon. The technical solution includes the following—
- Specific attack unit.
- Type of ordnance.
- Time of attack.
- Coordinating instructions.

Assess

1-35. The commander and staffers assess the results of mission execution. If combat assessment reveals that the commander's guidance has not been met, detect and deliver functions of the targeting process must continue to focus on the targets involved. This feedback may result in changes to original decisions made during the decide function. These changes influence the continue execution of the plan and made available to subordinate units as appropriate.

1-36. The targeting process is continuous and crucial to the synchronization of combat power. The identification and subsequent development of targets, the attack of the targets, and the combat assessment of the attacks provide the commander with vital feedback on the progress toward reaching the desired end state.

Targeting Interrelationships

1-37. While the targeting process may be labeled differently at the joint level the same targeting tasks are being accomplished, as demonstrated in table 1-1. For more information on the Joint Targeting Process see JP 3-60.

Table 1-1. Crosswalk of operations process, joint targeting cycle, D3A, and MDMP

Operations Process		Joint Targeting Cycle	D3A	MDMP	Targeting Task
Continuous Assessment	Planning	1. The End State and Commanders Objectives	Decide	Mission Analysis	• Perform target value analysis to develop fire support (including cyber/electromagnetic and inform/influence activities) high-value targets. • Provide fire support, inform/influence, and cyber/electromagnetic activities input to the commander's targeting guidance and desired effects.
		2. Target Development and Prioritization		Course of Action Development	• Designate potential high-payoff targets. • Deconflict and coordinate potential high-payoff targets. • Develop high-payoff target list. • Establish target selection standards. • Develop attack guidance matrix. • Develop fire support and cyber/electromagnetic activities tasks. • Develop associated measures of performance and measures of effectiveness.
		3. Capabilities Analysis		Course of Action Analysis	• Refine the high-payoff target list. • Refine target selection standards. • Refine the attack guidance matrix. • Refine fire support tasks. • Refine associated measures of performance and measures of effectiveness. • Develop the target synchronization matrix. • Draft airspace control means requests.
		4. Commander's Decision and Force Assignment		Orders Production	• Finalize the high-payoff target list. • Finalize target selection standards. • Finalize the attack guidance matrix. • Finalize the targeting synchronization matrix. • Finalize fire support tasks. • Finalize associated measures of performance and measures of effectiveness. • Submit information requirements to S-2.
	Preparation	5. Mission Planning and Force Execution	Detect		• Execute ISR Plan. • Update information requirements as they are answered. • Update the high-payoff target list, attack guidance matrix, and targeting synchronization matrix. • Update fire support and cyber/electromagnetic activities tasks. • Update associated measures of performance and measures of effectiveness.
	Execution		Deliver		• Execute fire support and electronic attacks in accordance with the attack guidance matrix and the targeting synchronization matrix.
		6. Assessment	Assess		• Assess task accomplishment (as determined by measures of performance). • Assess effects (as determined by measures of effectiveness).

SERVICE COMPONENTS TARGETING METHODOLOGY

1-38. Targeting occurs at all echelons within a joint command. Targeting is complicated by the requirement to deconflict procedures and priorities among the different Service components and multinational forces. The JFC is responsible for integrating attacks throughout the operational environment.

1-39. Each Service component has established unique doctrine and tactics, techniques, and procedures for targeting. Several publications address targeting procedures through their emerging doctrinal manuals. The habitual integration of resources from one or more Service components has been developed to support the targeting requirements of another Service component and multinational forces. Chapter 3 discusses the joint targeting process and phases.

1-40. Targeting is a multifaceted and a coordinated process. Existing Service components have four procedures in common for acquisition, selection, and attack of targets—
- Deciding in advance what is to be targeted.
- Locating the target.
- Attacking the target.
- Assessing the results of the attack.

1-41. This common approach to targeting mirrors the D3A methodology functions presented in this manual. The targeting process is accomplished by each Service components applying their developed tactics, techniques, and procedures within a joint framework established by the JFC. The organizational challenge for the JFC is to meld existing Service component architecture into an effective joint targeting working group for operational level targets without degrading their primary mission of targeting support to their respective components.

1-42. From the JFC perspective, a target is selected for strategic and/or operational reasons. Subsequently, a decision made whether to attack the target involves weapons employment. The targets selected or nominated in this process must support the JFC campaign plan and contribute to the success of present and future major operations. The JFC relies on the tactical level commanders to orchestrate the execution of matching the appropriate response to target. Control measures, such as a fire support coordination line, must be repositioned as needed to take full advantage of all assets available to the joint force. The JFC best influences the outcome of future tactical battles by setting the conditions for those battles and allocating resources to the Service components.

AIR-GROUND OPERATIONS RELATIONSHIP

1-43. The air and ground component commanders' have capabilities that overlaps The JFC assists in planning, coordinating, and integrating of these operations. Both components have intelligence collection assets and weapon systems with long-range capabilities. The capabilities of one Service component complement the capabilities of the other. Therefore, both air and ground weapons system must be synchronized to gain the greatest efficiency and technological advantage. This requires air and ground component commanders and their staffers to share the effort in acquiring and attacking targets throughout the operational area.

1-44. Staffers must understand the coordination requirements, measures to acquire, attacks targets safely and efficiently in an operational environment at all echelons. The battlefield is four-dimensional. The four dimensions are width, length, altitude or depth, and time. Current coordination and control measures for example fire support coordination measures (FSCM), airspace coordinating measures, and graphic/maneuver control measures permit the complementary, simultaneous attack of targets by air and ground weapons system.

TARGETING PERSONNEL AND RESPONSIBILITES

1-45. Key targeting working groups are members of the commander's coordinating and special staffs. These members perform the targeting process as part of their normal responsibilities within the MDMP. From their initial estimates and analysis, to their supervision and execution of the plan, they continue to revise and update their estimates. The relative formality of the decisionmaking process depends on time available and the level of the command (see FM 5-0).

1-46. The commander is responsible for the targeting effort. The intelligence, operations, and fire support officers form the core of the targeting working group at each level. The targeting working group has three primary functions in assisting the commander—
- Helps in synchronizing operations.
- Recommends targets to acquire and attack. The team also recommends the most efficient and available assets to detect and attack these targets.
- Identifies combat assessment requirements. Combat assessment can provide crucial and timely information to allow analysis of the success of the plan or to initiate revision of the plan. See Chapter 2 for more details on combat assessment.

1-47. The targeting effort is continuous at all levels of command. Continuity is achieved through parallel planning by targeting working groups from corps through battalion task force. Targeting is not just a wartime function. This process must be exercised before battle if it is to operate effectively. The members of the targeting working group must be familiar with their roles and the roles of the other team members. That familiarity can only be gained through staff training.

This page intentionally left blank.

Chapter 2
The Targeting Methodology

The modern battlefield presents high volume of targets and vulnerabilities for attack. The purpose of targeting methodology is to integrate and synchronize scalable fires with the maneuver operations. The targeting planning team has the responsibility to conduct planning, coordination, and deconfliction associated with the Army's targeting process. The purpose of this chapter is to explain the decide, detect, deliver, and assess (D3A) methodology, which is designed to enhance fire support planning and the intelligence targeting process.

GENERAL

2-1. Targeting is a combination of intelligence functions, planning battle command, weaponeering, operational execution, and combat assessment. Effective targeting identifies the targeting options, both lethal and nonlethal that support the commander's objectives. The D3A methodology facilitates the attack of the right target with the right asset at the right time (see figure 2-1).

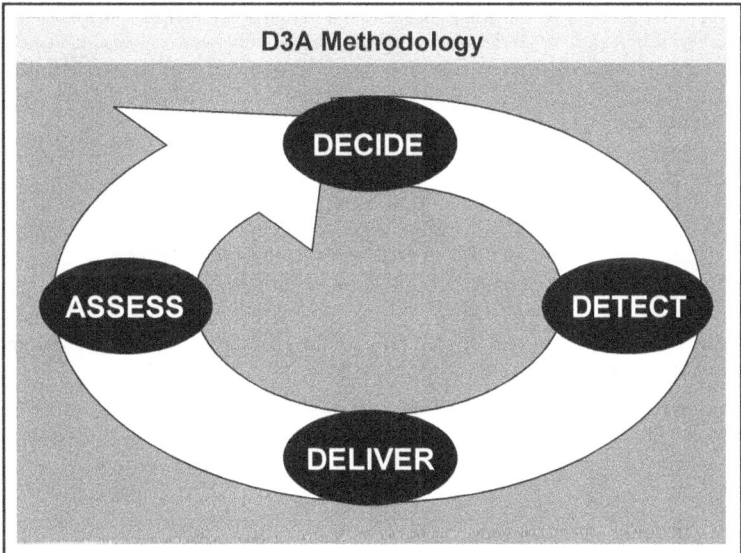

Figure 2-1. D3A methodology cycle

2-2. The targeting process provides an effective method for matching the friendly force capabilities against enemy targets. Lethal targets are best addressed with operations to kill, damage, disrupt, or capture; nonlethal targets are best engaged with civil-military operations, inform and influence activities, negotiation, political programs, economic programs, social programs, and other noncombat methods. In a counter insurgency operations, nonlethal targets are just as important as lethal targets and the targeting is frequently directed toward nonlethal options.

2-3. A very important part of the targeting process is the identification of potential fratricide situations and the necessary coordination measures to positively manage and control the attack of targets. These measures are incorporated in the coordinating instructions and appropriate annexes of the operation plan (OPLAN) and/or operation order (OPORD).

2-4. Targeting is a dynamic process. The process must keep up with the constant changing within area of operations. The tools and products described in this chapter must be updated based on combat assessment and situation understanding. Remember also, that the targeting process is repetitive. It is very seldom that decisions are made without any information from a previous targeting cycle. Intelligence from external agencies or intelligence previously generated internally feeds the decisionmaking.

2-5. The Army's targeting process comprises the following four functions—
- Decide which targets to engage.
- Detect the targets.
- Deliver (conduct the operation).
- Assess the effects of the operation.

DECIDE

2-6. The decide function is the first step in the targeting process. This step provides the overall focus and sets priorities for intelligence collection and attack planning. The decide functions draws heavily on a detailed intelligence preparation of the battlefield (IPB) and continuous assessment of the situation. Targeting priorities must be addressed for each phase or critical event of an operation. The decisions made are reflected in visual products. The products are as follow—
- The high-payoff target list (HPTL) is a prioritized list of high-payoff targets (HPT). The HPT is a target whose loss to the enemy and will significantly contribute to the success of the friendly course of action (COA). HPT is those high-value targets (HVT) that must be acquired and successfully attacked for the success of the friendly commander's mission. The HVT is a target the enemy commander requires for the successful completion of the mission. The loss of a HVT is expected to degrade important enemy functions significantly throughout the friendly commander's area of interest.
- The intelligence, surveillance, and reconnaissance (ISR) plan is designed to answer some of the commander's priority intelligence requirements (PIR), to include those HPT designated as PIR. The plan, within the availability of additional collection assets, supports the acquisition of more HPT. Determining the intelligence requirements is the first step in the collection management process. For a more detailed description. See Field Manual (FM) 5-0.
- Target selection standards (TSS) address accuracy or other specific criteria that must be met before targets can be attacked.
- **The *attack guidance matrix* (AGM) is a matrix, approved by the commander, which addresses which targets will be attacked, how, when, and the desired effects.**

2-7. The products of the decide function are briefed to the commander. Upon his approval, his decisions are translated into the OPORD with annexes. Specific targeting products are required at echelons indicated in figure 2-2 below.

Product	Corps	Division	Brigade Combat Team	Task Force
HPTL	Yes	Yes	Yes	Yes[1]
ISR Plan	Yes	Yes	Yes	
TSS	Yes	Yes	Yes	Yes[1]
AGM[2]	Yes	Yes	Yes	Yes[1]

[1] At battalion task force level, the high-payoff target list, attack guidance matrix, and target selection standards should be addressed. The process is very informal and may not result in written products. The products produced by the brigade may be used by the battalion task force.

[2] At brigade level and below, the fire support execution matrix provides attack guidance. The fire support execution matrix is discussed in detail in FM 6-20-40 and FM 6-20-50, to be revised and renumbered as FM 3-09.42.

Legend:
AGM – attack guidance matrix
ISR – intelligence, surveillance and reconnaissance
HPTL – high- payoff target list
TSS – target selection standards

Figure 2-2. Targeting products (example)

MISSION ANALYSIS

2-8. The commander and staff plans for future operations by analyzing one or more alternative COAs at all echelons. Each COA is based on the following—
- Mission analysis.
- Current and projected battle situations.
- Exploit opportunities.

2-9. The process begins with receipt of a mission, whether assigned by higher headquarters or deduced by the commander. The commander, with or without input from his staff, analyzes the mission; considers tasks that must be performed, their purpose, and limitations on the unit. The completed analysis is the basis for developing a restated mission. The restated mission is the origin from which to start the targeting process.

INTELLIGENCE PREPARATION OF THE BATTLEFIELD

2-10. IPB can be best described as the process of understanding the battlefield, the enemy, and the options presented. The objective of IPB is the early identification of probable enemy COA. It is a continuous and systematic method for analyzing the enemy, weather, and terrain in a geographical area. The IPB provides much of the information for the intelligence estimate and is the foundation for the rest of the targeting process.

2-11. IPB begins with analysis of enemy doctrine and capabilities in a geographical area, the terrain, and the effects of weather on that terrain to include tactics, techniques, and procedures that threat/adversary forces prefer to employ. The products of IPB are HVT list, civil considerations, modified combined obstacle overlay, threat models, threat capabilities, weather effects matrix, and event template/matrix. These products are used to visualize the threat/adversary characteristics, predict enemy intentions, and develop COA with statements. These products assist in target value analysis and initial identification of potential HVT. Doctrinal templates convert the threat characteristics into graphics. Situation graphics help in refining HVT for specific area of operations and threat/adversary COA.

2-12. Concurrent with development of the situation graphics are an examination of enemy decision points and/or critical nodes as a part of each COA. The examination shows what might happen if the enemy commander's plan fails and what actions make up his failure options. Evaluation of threat/adversary COA with statement leads to identification of critical enemy functions in each COA and the HVT associated with each function.

2-13. IPB in counterinsurgency operations places greater emphasis on civil considerations, especially to win the local population's support to defeat an insurgency in the area of operations. A continuous IPB process identifies new intelligence requirements. The IPB products are revised throughout the operation. Intelligence and civil affairs personnel provide information on the relative importance of different target personalities and areas and the projected effects of scalable fires. Specifically, the intelligence analysts need to identify individuals and groups to engage as potential counterinsurgency supporters, targets to isolate from the population, and targets to eliminate. IPB in counterinsurgency operations requires personnel to work in areas like economics, anthropology, and governance that may be outside their expertise. Civil affairs forces are subject-matter experts. The personnel trained and equipped specifically to analyze the civil aspects of the common operating picture. Therefore, integrating staffs and drawing on the knowledge of personnel not trained in intelligence operations and external experts with local, regional, and cultural knowledge are critical to effective preparation.

2-14. Applying the IPB process helps the commander to selectively apply and maximize combat power at critical points in time and space. It does this by describing the battlefield environment, how the natural environment affects friendly units and likely threat/adversary COAs. Situational graphics support the development of event templates. Event templates help identify critical enemy activities. It also identifies named areas of interest where specific enemy activities or events will help confirm or deny the adoption of a particular COA. Potential HPTs are identified. HPTs are those HVT that must be attacked to give the commander a significant advantage in defeating the enemy. This work is further analyzed in the war gaming process.

> *Note.* See the Field Manual Interim (FMI) 2-01.301 and FM 3-24 for additional details.

TARGET VALUE ANALYSIS AND WAR GAMING

2-15. The intelligence staff and targeting officer evaluates and integrates the various factors of the operational environment that affects both friendly and threat/adversary operations. This coordination helps to develop the intelligence summary which contains HPT, HVT lists, and may include high-value individual (HVI). Target value analysis yields HVT for a specific enemy COA. Target value analysis involves a detailed analysis of enemy doctrine, tactics, equipment, organizations, and expected behavior for a selected COA. The target value analysis process identifies potential HVT sets associated with critical enemy functions that could interfere with the friendly COA or that are vital to enemy success.

2-16. Target spreadsheets (or target folders, as appropriate) identify the HVT in relation to a type of operation. The target spreadsheets give detailed targeting information for each HVT. The information on target spreadsheets and target sheets are used during the IPB and the war gaming processes. The targeting section within the analysis and control element under the operational control of the assistant chief of staff, intelligence (G-2) incorporates all-source intelligence to develop both tools.

2-17. The target value analysis process provides a relative ranking of target sets. The war game begins when the target analyst in the G-2 or the intelligence staff officer (S-2) plays the role of the threat commander or acts as the ISR officer.

2-18. Alternative friendly developments are analyzed in terms of their impact on enemy operations and likely responses during war gaming. The enemy battlefield functions that must be attacked to force the best enemy response are identified. The commander and his staff analyze the criticality of friendly battlefield functions with regard to a specific COA. The best places to attack HPT become more refined during war gaming of friendly options. These places are called target areas of interest. Target areas of interest are points or areas where the commander can acquire and engage HPTs by fires and/or maneuver. Decision points or decision time phase lines are used to ensure that the decision to engage or not to engage occurs at the proper time. Decision points and target areas of interest are recorded on the assistant chief of staff,

operations' (G-3) decision support template. The purpose of war gaming is to finalize individual running estimates and to develop all of the following—
- Scheme of maneuver.
- Fire support plan.
- Friendly decision support template.

2-19. HVTs are identified and prioritized during the war gaming phase of planning. In addition, it identifies the subset of HVTs that must be acquired and attacked for the friendly mission to succeed. HVTs may be nominated as HPTs when these targets can be successfully acquired, vulnerable to attacks, and such an attack supports the commander's scheme of maneuver. Once identified and nominated, HPTs are grouped into a list identifying them for a specific point in the battle. The completed HPTL is submitted to the commander for approval. The approved HPTL becomes a formal part of the fire support plan.

2-20. The G-3 or operations staff officer (S-3) normally leads the war game and role-plays the friendly COA statements. The operations section establishes the technique and recording method for the war game. The G-2 or S-2 role plays the enemy's most dangerous and likely COA by using enemy doctrine and tactics. When available, the assistant chief of staff, and plans/plans staff officer war games the civilian COA. The deputy chief of fires or fire support officer (FSO) advises the G-3/S-3 on using available fire support weapons system and records the needs for fire support. The G-3/S-3 uses the war game to determine adequacy of fire support. Operations staff works directly with the intelligence section to ensure full use of fire support target acquisition assets in the intelligence collection plan. The air liaison officer and deputy or assistant aviation officer advises on availability and employment of air assets to include airspace coordinating measures. Other key staff officers who are vital and should be consulted, consist of the following—
- G-3 air or brigade aviation officer for airspace integration with organic and supporting airspace users.
- Logistics officer for supportability considerations.
- Engineer officer for mobility, countermobility, survivability, and environmental considerations.
- Air defense artillery officer for air defense coverage.

2-21. Input from the rest of the staff achieves a complete analysis of the impact of all warfighting functions. This ensures the AGM is synchronized with the decision support template, and selection of HPT is supported by PIR and the intelligence collection tasks.

TARGET SELECTION

2-22. The staff war games different COA statements to develop the HPT. As each friendly option is war gamed by the staff, the G-2 or S-2 identifies HVT from which the staff nominates HPT. The HPT is targets that are critical to friendly success. Targets that can be acquired and attacked are candidates for the HPTL. Targets that need outside acquisition or attack are sent to higher headquarters. The key to HPT is that they are critical to the enemy commander's needs (HVT) and the friendly concept of the operation. HPTs support the friendly force commander's scheme of maneuver and intent. The war game phase helps the commander to focus reconnaissance assets on HPTs to conduct battle damage assessment (BDA). The analysis and control element's collection manager helps identify and task the sensors needed for collection of the HPT. The collection manager can determine the best sensor and its availability by referencing the ISR synchronization matrix. A detailed discussion of the ISR synchronization matrix is contained in TC 2-01.

HIGH-PAYOFF TARGET LIST

2-23. The HPTL identifies the HPTs by phases in the battle and order of priority in the figure 2-3. Target value is usually the greatest factor contributing to target payoff. However, other things to be considered include the following—
- The sequence or order of appearance.
- The ability to detect, identify, classify, locate, and track the target. (This decision must include sensor availability and processing timeline considerations.)

Chapter 2

- The degree of accuracy available from the acquisition system(s).
- The ability to engage the target.
- The ability to defeat the target on the basis of attack guidance.
- The resources required to do all of the above.

Phase of the Operation—1—Isolate the Enemy Unit:		
Priority	Category	High-payoff target
1	Fire Support	Mortars
2	Maneuver	Insurgent teams
3	Electromagnetic Activities	Cell phone
4	Electromagnetic Activities	FM radio
5	Civilians	Hostile civilian crowds
*A hostile crowd is defined as 25 or more people with leadership interfering or capable of interfering with the BCT operations.		
Legend: BCT – brigade combat team FM – frequency modulation		

Figure 2-3. High-payoff target list (example)

2-24. Targets are prioritized according to the considerations above within specific time windows. The targeting working group sets priorities for the targets according to its judgment and the advice of the fires cell targeting officer and the field artillery intelligence officer (FAIO). Target spreadsheets give a recommended priority and attack sequence. If the target spreadsheet or war gaming departs from the commander's guidance, it is noted on the proposed HPTL to inform the commander of the conflict. The target category of the HPT is shown, either by name or by number, on the list. The category name and number are shown on the target spreadsheet. The number of target priorities should not be excessive. Too many priorities will dilute the intelligence collection acquisition and attack efforts. The approved list is given to the operations, intelligence, and fires cells. It is used as a planning tool to determine attack guidance and to refine the collection plan. This list may also indicate the commander's operational need for BDA of the specific target and the time window for collecting and reporting it.

Note. Any format serves the purpose of a HPTL for linking targets with phases of battle.

2-25. One way to organize the HPTL is to group all HPTs into target sets that reflect the capabilities and functions the commander has decided to engage. Target sets are identified and prioritized for each phase of the operation. Within the sets, individual targets are rank ordered by target value, sequence of appearance, importance, or other criteria that satisfy the commander's desired effects. In this way, the targeting working group reduces, modifies, and reprioritizes HVTs while ensuring that HPTs support the concept of operations.

2-26. The commander's guidance may require changes, which should be annotated on the HPTL. The target name or number and description are placed on the list for specific HPTs in each category. Once the commander approves or amends the HPTL, it goes back to the targeting working group to help them develop the AGM and collection plan.

INTELLIGENCE, SURVEILLANCE, AND RECONNAISSANCE PLAN

2-27. The G-2/S-2 develops collection strategies that support the commander's concept of operations with available resources. Collection management orchestrates the intelligence system of weapons system to focus the intelligence effort in support of operations. If BDA is needed, collection is planned to satisfy that requirement as well.

2-28. ISR is a continuous combined arms effort led by the operations and intelligence staffs in coordination with the staff that sets reconnaissance and surveillance in motion. The PIR and other intelligence requirements drive the collection effort. The commander takes every opportunity to improve his situational understanding about the enemy and terrain. Commanders integrate reconnaissance and surveillance to form an integrated ISR plan that capitalizes on their different capabilities. The ISR plan is often the most important part of providing information and intelligence that contribute to answering the commander's critical information requirement. For the G-2/S-2, an effective ISR plan is critical in answering the PIR. Upon the completion of planning, the initial ISR plan becomes annex L (Intelligence, Surveillance, and Reconnaissance) of the OPORD/OPLAN. See FM 5-0 for additional information.

2-29. The ISR plan is not a military intelligence specific product—although the G-3/S-3 is the staff proponent of the ISR plan—it is an integrated staff product executed by the unit at the direction of the commander. The G-2/S-2, however, must maintain situational understanding in order to recommend changes or further development of the ISR plan. Based on the initial IPB and commander's critical information requirement, the staff—primarily the G-2/S-2—identifies gaps in the intelligence effort and develops an initial ISR plan based on available ISR assets. The G-3/S-3 turns this into an initial ISR annex that tasks reconnaissance and surveillance assets as soon as possible to begin the collection effort.

2-30. The G-3/S-3, assisted by the G-2/S-2, uses the ISR plan to task and direct the available ISR assets to answer the PIR and intelligence requirements. Conversely, the staff revises the plan as other intelligence gaps are identified if the information is required to fulfill the commander's critical information requirement or in anticipation of future intelligence requirements. With staff participation, the G-2/S-2 intelligence officer synchronizes the collection effort through a complementary product to the ISR plan - the intelligence synchronization plan.

TARGET SELECTION STANDARDS

2-31. **Target selection standards (TSS) are criteria applied to enemy activity (acquisitions and battlefield information) and used in deciding whether the activity is a target.** TSS put nominations into two categories targets and suspected targets. Targets meet accuracy and timeliness requirements for attack. Suspected targets must be confirmed before any attack. See Annex D for a sample target selection standards worksheet. Units may develop their own worksheet format.

2-32. TSS is based on the enemy activity under consideration and available weapon systems by using the following—

- Weapon system target location accuracy requirements (target location error [TLE]). Special consideration must be given to TLE for the employment of guided precision munitions.
- Size of the enemy activity (point or area).
- Status of the activity (moving or stationary).
- Timeliness of the information,

2-33. Considering these factors, different TSS may exist for a given enemy activity based on different weapons system. For example, an enemy artillery battery may have a 150-meter TLE requirement for attack by cannon artillery and a 1 kilometer requirement for attack aircrafts. TSS is developed by the fires cell in conjunction with the military intelligence personnel. Intelligence analysts use TSS to quickly determine targets from battlefield information and pass the targets to the fires cell. Weapon system managers such as fires cells, fire control elements, or fire direction centers use the TSS to identify targets for attack quickly. Commands can develop standard TSS based on threat characteristics and doctrine matched with the standard available weapon systems.

2-34. TSS worksheet is given to the G-2 or S-2 by the fires cell. The FAIO use TSS to identify targets that are forwarded to a fires cell. Intelligence analysts evaluate the source of the information as to its reliability

Chapter 2

and accuracy, confirm that the size and status of the activity meet the TSS, and then compare the time of acquisition with the dwell time. Accurate information from a reliable source must be verified before declaring it a target if the elapsed time exceeds dwell time.

Note. Dwell time is the length of time a target is doctrinally expected to remain in one location.

2-35. The G-2 or S-2 knows the accuracy of acquisition systems, associated TLE, and the expected dwell times of enemy targets. He can then specify whether information he reports to the weapon system manager is a target or a suspected target. Some situations require intelligence assets to identify friendly or foe before approval to fire is given. HPT that meet all the criteria should be tracked until they are attacked in accordance with the AGM. Location of targets that do not meet TSS should be confirmed before they are attacked. The TSS can be graphically depicted in a TSS matrix as shown in the figure 2-4.

High-payoff target	Timelines	Accuracy
Mortars	10 minutes	100 meters
Insurgent teams	30 minutes	100 meters
Cell phone	Within two hours of H-hour	Placed/received within 12 km of Fustina airfield
FM radio	20 minutes	150 meters
Hostile civilian crowds	Within six hours of H-hour	Within 12 km of Fustina village

Legend: FM – frequency modulation   H - hour (H-hour is the time for a scheduled event to begin.)

Figure 2-4. Target selection standards matrix (example)

2-36. The matrix lists each weapon system that forwards targets directly to the fires cell, fire control element, or fire direction center. The effects of weather and terrain on the collection assets and on enemy equipment are considered. TSS is keyed to the situation. However, the greatest emphasis is on the enemy situation, considering deception and the reliability of the source or agency that is reporting.

ATTACK GUIDANCE

2-37. Knowing target vulnerabilities and analyzing the probable effect an attack will have on enemy operations allows a staff to propose the most efficient available attack option. Key guidance is whether the commander wishes to disrupt delay, limit damage, or destroy the enemy. During war gaming, decision points linked to events, areas of interest, or points on the battlefield are developed. These decision points cue the command decisions and staff actions where tactical decisions are needed.

2-38. Based on commander's guidance, the targeting working group recommends how each target should be engaged in terms of the effects of fire and attack options to use. Effects of fire can be to harass, suppress, neutralize, or destroy the target. The subjective nature of what is meant by these terms means the commander must ensure the targeting working group understands his use of them. Application of fire support automation system default values further complicates this understanding.

2-39. The decision of what weapon system to use is made at the same time as the decision on when to acquire and attack the target. Coordination is required when deciding to attack with two different means such as electronic warfare (EW) and combat air operations. Coordination requirements are recorded during the war game process.

2-40. The commander, with recommendations from the targeting working group, must approve the attack guidance. This guidance should detail the following—
- A prioritized list of HPT.
- When, how, and desired effects of attack.

The Targeting Methodology

- Any special instructions.
- HPTs that require BDA.

2-41. This information is developed during the war game. Attack guidance applies to both planned targets and targets of opportunity. Accordingly, attack guidance may address specific or general target descriptions. Attack guidance is provided to weapon system managers via the AGM.

ATTACK GUIDANCE MATRIX

2-42. The AGM consists of columns for the following—
- Specific HPT.
- Timing of attack.
- How targets are attacked.
- Target categories.
- Restrictions.

Note. An example of the AGM is shown in figure 2-5.

High-payoff target	When	How	Effect	Remarks
Mortars	I	Field Artillery	Destroy	Use search and attack teams in restricted areas
Insurgent teams	I	Field Artillery	Neutralize	Destroy mission command
Cell phone	A	Electronic Attack	Disrupt	Disrupt service starting H-2
FM radio	A	Electronic Attack	Disrupt	No jamming until H-3 to preserve intelligence
Hostile civilian crowds	A	MISOP/MP	Dispersed	25 or more with leadership constitute crowd

Legend:
(A) - as acquired FM – frequency modulation H - hour (H-hour is the time for a scheduled event to begin.)
(I) – immediate and special case MISOP – military information support operations MP – military police

Figure 2-5. Attack guidance matrix (example)

High-Payoff Target Column

2-43. This column lists the prioritized HPTs identified during war gaming. These targets have priority for engagement.

WHEN Column

2-44. Timing the attack of targets is critical to maximizing the effects. During war gaming, the optimum time is identified and reflected in the WHEN column. The letter P indicates that the target should not be engaged now but should be planned for future firing (for example, a preparation, a suppression of enemy air defense (SEAD) program, or a countermobility program) or simply should be put on file. Such targets should be engaged in the sequence that they are received in the headquarters, with respect to the priority noted in the HPTL. Designators (A, I, and P) on should be limited to a very small percentage of targets and only for the most critical types. Too many immediate targets are disruptive and lower the efficiency of weapon systems. Immediate attacks take precedence over all others and are conducted even if weapon systems must be diverted from attacks already underway. Some examples of very important targets include—
- Missile systems capable of chemical, biological, radiological, and nuclear attack.
- Division headquarters.
- Certain identified individuals.
- Chemical, biological, radiological, and nuclear weapons storage and support facilities.

2-45. The G-3 or S-3 and chief of fires/brigade FSO must establish procedures within the main command post (CP) that allow for immediate attack of targets.

HOW Column

2-46. The HOW column links the weapon system to the HPT. It is best to identify a primary and backup weapon system for attack of HPTs.

EFFECTS Column

2-47. Effects refer to the target attack criteria. The targeting working group should specify attack criteria according to the commander's general guidance. Target attack criteria should be given in quantifiable terms (for example, as a percentage of casualties or destroyed elements, time, ordnance, and allocation or application of assets). In addition, it can be noted as the number of battery or battalion volleys.

REMARKS Column

2-48. Some examples of how this column should be used are—
- Note accuracy or time constraints.
- Note required coordination.
- Limitations on the amount or type of ammunition.
- Any need for BDA.

2-49. This column should note which targets should not be attacked in certain tactical situations (for example, targets not to be attacked if the enemy is withdrawing).

2-50. As the operation progresses through time, the AGM may change. The AGM is a tool that must be updated based on the changing enemy situation. It should be discussed and updated during routine staff planning meetings. Consider separate AGM for each phase of the concept of operations.

FORMATS

2-51. The formats for the HPTL, TSS, and AGM presented in the preceding paragraphs are examples only. Targeting personnel must understand all the considerations that are involved in building these targeting tools. However, experienced staffs may prefer to develop their own formats tailored for their situation. Alternative formats are provided in Appendix D.

DETECT

2-52. Detect is the next critical function in the targeting process. The G-2 or S-2 is responsible for directing the effort to detect HPTs identified in the decide function. The ability to identify the specific who, what, when, and how for target acquisition, the G-2 or S-2 must work closely with all of the following—
- Analysis and control element.
- Assistant chief of staff, information engagement.
- Information engagement staff officer (S-7).
- FAIO.
- Targeting officer and/or FSO.

2-53. This process determines accurate, identifiable, and timely requirements for collection systems. The analysis and control element's targeting section is responsible for ensuring that the collection system asset managers understand these requirements.

2-54. Information needs for target detection are expressed as PIR and/or information requirements. Their relative priority depends on the importance of the target to the friendly scheme of maneuver and tracking requirements coupled with the commander's intent. PIRs support detection of HPT incorporated into the overall collection plan of the unit.

2-55. Targets are detected and tracked by the maximum use of all available assets. The G-2 or S-2 must focus the intelligence acquisition efforts on the designated HPTs and PIRs. The collection manager considers the availability and capability of all collection assets at the strategic, operational, and tactical levels. The joint force assets are available to the collection manager. The intelligence officer translates the PIR and intelligence requirement into specific information requirements and specific orders and requests. If possible, he arranges direct dissemination of targeting information from the collector to the targeting cell or targeting intelligence to the fires cell.

2-56. In counterinsurgencies, intelligence regarding factors of the operational environment affecting the populace requires particular attention. Such intelligence is important for developing political, social, and economic programs. Intelligence personnel continuously analyze large quantities of all-source intelligence reporting to determine the following—

- Threat validity.
- Actual importance of potential targets.
- Best means to engage the target.
- Expected effects of engaging the targets (which will guide actions to mitigate negative effects).
- Any changes required to the exploitation plan.

DETECTION PROCEDURES

2-57. It is essential that all ISR assets be used effectively and efficiently. Duplication of effort among available assets must be avoided unless it is required to confirm target information. The intelligence cell develops and manages the collection plan to avoid duplication at corps and division level. At the same time, the intelligence cell ensures that no gaps in planned collection exist. This allows timely combat information to be collected to answer the commander's intelligence requirements. This information lets analysts develop the enemy situation and identify targets.

2-58. Desired HPTs must be detected in a timely, accurate manner. Clear and concise tasks must be given to the reconnaissance units or surveillance systems that can detect a given target. Mobile HPTs must be detected and tracked to maintain a current target location. Target tracking is inherent to detection. The fires cell tells the G-2 or S-2 the degree of accuracy required and dwell time for a target to be eligible for attack. The G-2 or S-2 must match accuracy requirements to the TLE of the collection systems. If the target type and its associated signatures (electronic, visual, thermal, and so forth) are known, the most capable collection asset can be directed against the target. The asset can be placed in the best position according to estimates of when and where the enemy target will be located.

2-59. As the assets collect information for target development, it is forwarded to the intelligence analysts of the analysis and control element. They use the information in performing situation and target development. When the analysts identify a target specified for attack, it is passed to the fires cell. The fires cell executes the attack guidance against the target. Close coordination among the intelligence staff and the fires cell is essential to ensure that the targets are passed to a weapon system that will engage the target. To ensure the exchange is timely, the FAIO must have access to the analysis and control element workstation. The FAIO coordinates with the G-2 and fires cell to pass HPTs and other targets directly to the fire control element at the fires battalion or fires brigade or, if approved by the maneuver commander, directly to a firing unit. The result is an efficient attack of targets that have been designated in advance for attack. Some units have found it advantageous to locate the FAIO in the analysis and control element with communications to the fires cell. The FAIO notifies the fires cell immediately when intelligence information warrants attack. This allows the FAIO to focus on intelligence information analysis and the fires cell to manage the control of fires. The FAIO functions are performed by the targeting officer at brigade and the battalion's S-2.

2-60. Tracking is an essential element of the detect function of the targeting process. Tracking priorities are based on the commander's concept of the operation and targeting priorities. Tracking is executed through the collection plan. Not all targets will be tracked. However, many critical targets move frequently or constantly. As such, these HPTs require tracking.

THE ISR SYNCHRONIZATION MATRIX

2-61. The ISR synchronization matrix is a product used by the intelligence officer to ensure that collection tasks are tied to scheme of maneuver in time and space, effectively linking reconnaissance and surveillance to maneuver and effects. The ISR synchronization matrix is typically constructed in spreadsheet format and is always accompanied by an ISR overlay that graphically depicts the information contained in the matrix. The intelligence officer uses the matrix to synchronize reconnaissance and surveillance tasks in the same way the operations officer uses the maneuver synchronization matrix to synchronize the overall unit scheme of maneuver.

INTELLIGENCE SYNCHRONIZATION

2-62. The intelligence officer, with staff participation, orchestrates the entire collection effort to include all assets the commander controls, assets of lateral units and higher echelon units and organizations, and intelligence reach to answer the PIR and other intelligence requirements. Intelligence synchronization activities include the following—

- Conducting requirements management includes—
    - Anticipate.
    - Develop.
    - Analyze.
    - Validate.
    - Prioritize intelligence requirements.
- Recommending PIR to the commander. Manage the commander's intelligence requirements, requests from subordinate and lateral organizations, and tasks from higher headquarters. Eliminate satisfied requirements and add new requirements as necessary.
- Developing specific information requirements from the PIR.
- Converting the specific information requirements into reconnaissance and surveillance tasks. The S-2/G-2 assigns intelligence production and tasks to subordinate intelligence elements or personnel, submits requests for information to higher and lateral echelons, and coordinates with (or assists) the G-3/S-3 to develop and assign reconnaissance and surveillance tasks.
- Comparing the reconnaissance and surveillance tasks to the capabilities and limitations of the available ISR assets (in coordination with the operations officer).
- Forwarding specific information requirements that cannot be answered by available assets to higher or lateral organizations as requests for information.
- Assessing collection asset reporting and intelligence production to evaluate the effectiveness of the collection effort.
- Maintaining situational understanding to identify gaps in coverage and to identify the need to redirect ISR assets.
- Updating the ISR plan. The G-2/S-2 manages and updates the ISR plan, and ISR synchronization matrix, as PIR is answered and new requirements arise.

ESSENTIAL TARGET INFORMATION

2-63. Targets and suspected targets may be passed to the targeting working group by a number of means. It is important that the essential information be passed for proper analysis and attack to take place. As a minimum, the target report must include the following—

- Reporting agency.
- Date time group of acquisition by the sensor.
- Description of the activity.
- Size of the target.
- Target location and altitude.
- TLE.
- Dwell time.
- Status (stationary or moving).

2-64. The date time group is important as the dwell time of the target is analyzed. The dwell time of the target determines whether to attack based on the likelihood of the target having moved.

2-65. The target description and size are compared with the AGM. Description should include posture (dug in or in the open) and activity (moving or stationary). This information is used to determine the following—
- Attack means.
- Intensity of attack.
- Number of assets to be committed.
- Other technical considerations.

2-66. The target location must be given as accurately as possible within the confines of timeliness. The targeting working group can request TLE for a target on the basis of the attack criteria. However, a sensor may report a target with a large TLE. The target will still be processed and the team will determine whether to engage the target and by what means.

2-67. Figure 2-6 provides a blank format of a sample target report. For a filled in sample see Annex D. Units may develop their own format.

LINE NUMBER

1. Report Agency: _____

2. Type of Sensor: _____

3. Report DTG: _____

4. Acquisition DTG: _____

5. Distribution: _____

6. Posture[1]: _____

7. Activity[2]: _____

8. Size[3]: _____

9. Location[4]: _____

10. Location Error[5]: _____

NOTES:

[1]Dug-in, in the open, in built-up areas, and so on.

[2]Moving (direction) or stationary.

[3]Unit size, diameter, and so on.

[4]Grid coordinates

[5]+ / - meters

Figure 2-6. Target report

TARGET DEVELOPMENT

2-68. *Target development* is the systematic examination of potential target systems and their components, individual targets, and even elements of targets to determine the necessary type and duration of the action that must be exerted on each target to create an effect that is consistent with the commander's specific objectives (Joint Publication [JP] 3-60). Target development includes functions such as target research, nomination, deconfliction, aimpoint recommendation, target materials production, and collateral damage estimation. Target development generally results in four products: target development nominations, target folders, collection and exploitation requirements, and target briefs. Detailed analysis should characterize the function, criticality, and vulnerabilities of each potential target, linking targets back to targeting objectives and measures of effectiveness.

TARGET VETTING

2-69. *Vetting* is a part of target development that assesses the accuracy of the supporting intelligence to targeting (JP 3-60). It is a key component of the target development process to establish a reasonable level of confidence in a candidate target's functional characterization. In target vetting, G-2/S-2 coordinates an intelligence community review of the target data for accuracy of the supporting intelligence. An assessment of the supporting intelligence will include a minimum of target identification, significance, collateral data estimation, geospatial or location issues, impact on the enemy or friendly forces, impact of not conducting operations against the target, environmental sensitivity, and intelligence gain/loss concerns. Vetting does not include an assessment of compliance with the law of war or rules of engagement.

TARGET VALIDATION

2-70. *Validation* is a part of target development that ensures all vetted targets meet the objectives and criteria outlined in the commander's guidance and ensures compliance with the law of war and rules of engagement (JP 3-60). Targets are validated against multinational concerns in a bilateral environment. Target vetting and validation should be revisited as new intelligence becomes available or the situation changes. Target validation is done by targeteers, in consultation with the planners and other experts/agencies, as required.

2-71. Target validation asks such questions as—
- Does the desired target effect contribute to achieving one or more of the commander's objectives, achieve desired operational effects, or achieve supporting sub tasks?
- Does the desired target effect support the end state?
- Does the desired target effect comply with the commander's guidance and intent?
- Is attacking the target lawful? What is the law of war and rules of engagement considerations?
- Does the target contribute to the adversary's capability and will to wage war?
- Is the target (still) operational? Is it (still) a viable element of a target system? Where is the target located?
- Will striking the target arouse political or cultural "sensitivities"?
- How will striking the target affect public opinion (enemy, friendly, and neutral)?
- Are there any facilities or targets on the no-strike list or restricted target list collocated with the target being validated?
- What is the relative potential for collateral damage or collateral effects, to include casualties? Consider collateral damage concerns in relation to law of war, rules of engagement, and commander's guidance.
- What psychological impact will operations against the target have on the adversary, friendly forces, or multinational partners?
- What would be the impact of not conducting operations against the target?
- Is it feasible to attack this target at this time? If not, could it be targeted at another time? What is the risk?
- Would attacking the target generate significant environmental impacts or arouse environmental sensitivities?
- Will attacking the target negatively affect friendly operations due to current or planned friendly exploitation of the target?
- How will actions taken against the target impact on other operations?

● What is the target's proximity to friendly elements?

2-72. Figure 2-7 provides several examples of both desirable and undesirable effects to be considered during target validation.

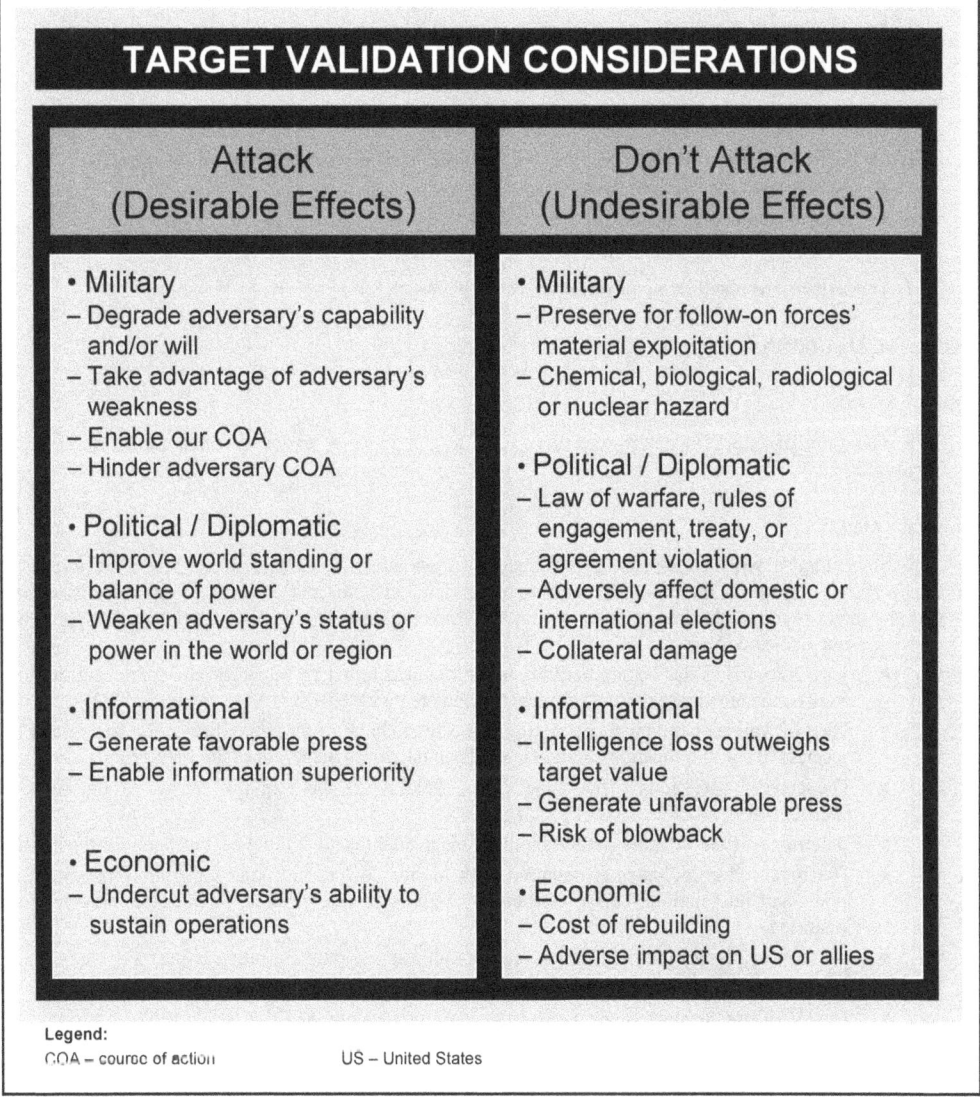

Figure 2-7. Target validation considerations

DELIVER

2-73. The deliver function of the targeting process executes the target attack guidance and supports the commander's battle plan once the HPT has been located and identified.

Chapter 2

ATTACK OF TARGETS

2-74. The attack of targets must satisfy the attack guidance developed in the decide function. Target attack requires several decisions and actions. These decisions fall into two categories: tactical and technical.

2-75. Tactical decisions determine the following—
- The time of the attack.
- The desired effect, degree of damage, or both.
- The weapon system to be used.
- Potential for collateral damage or collateral effects.

2-76. On the basis of these tactical decisions, the technical decisions describe the following—
- Number and type of munitions.
- Unit to conduct the attack.
- Response time of the attacking unit.

2-77. These decisions result in the physical attack of the target.

TACTICAL DECISIONS

Time of Attack

2-78. The time of attack is determined according to the type of target: planned target or target of opportunity.

Planned Targets

2-79. Some targets will not appear as anticipated. Target attack takes place only when the forecasted enemy activity occurs in the projected time or place. The detection and tracking of activities associated with the target becomes the trigger for target attack. Once the designated activity is detected the targeting working group does the following—
- The G-2 verifies the enemy activity as the planned target to be attacked. Monitoring decision points and named/target areas of interest associated with HPTs.
- The G-2 validates the target by conducting a final check of the reliability of the source and the accuracy (time and location) of the target. Then he passes the target to the fires cell.
- The current operations officer checks the legality of the target in terms of the rules of engagement.
- Determines if the weapon system(s) planned is available and still the best weapon for the attack.
- The fires cell coordinates as required with higher, lower, and adjacent units, other Services, allies, and host nation. This is particularly important where potential fratricide situations are identified.
- The fires cell issues the fire mission request to the appropriate executing unit(s).
- The fires cell informs the G-2 of target attack.
- The G-2 alerts the appropriate assessment asset responsible for BDA (when applicable).

Targets of Opportunity

2-80. HPTs of opportunity are processed the same as planned HPTs. Targets of opportunity not on the HPTL are first evaluated to determine when, or if, they should be attacked. The decision to attack targets of opportunity follows the attack guidance and is based on a number of factors such as the following—
- Activity of the target.
- Dwell time.
- Target payoff compared to other targets currently being processed for engagement.

2-81. If the decision is made to attack immediately, the target is processed further. The availability and capabilities of weapon systems to engage the target are assessed. If the target exceeds the capabilities or availability of the unit weapon systems, the target should be sent to a higher headquarters for immediate attack. If the decision is to defer the attack, continue tracking, determine decision point(s) for attack, and modify collection tasking as appropriate.

Desired Effects

2-82. *Desired effects* **achieve damage or casualties to the enemy or material that a commander desires to achieve from an identical target engagement. Damage effects on material are classified as light, moderate, or severe.**

2-83. Effects of fires can only be properly assessed by an observer or with an analyst. It is important that each target has a primary and alternate observer at the brigade combat team (BCT) and task force level. Each observer must understand the desired effects to include the when and for how long they are required. Emphasis on this issue during training will enhance the effectiveness and efficiency of fire support.

Weapons System

2-84. The last tactical decision to be made is the selection of the appropriate weapon system. For planned targets, this decision should have been made during the decide function of the targeting process. A check must be made to ensure that the selected weapon system is available and can conduct the attack. If not, the targeting working group must determine the best weapons system available to attack the target.

2-85. A key part of determining the appropriate method of attack is weaponeering. *Weaponeering* is defined as the process of determining the quantity of a specific type of lethal or nonlethal weapons required to achieve a specific level of damage to a given target, considering target vulnerability, weapons characteristics and effects, and delivery parameters (JP 3-60). Weaponeering also considers such things as enemy actions (the effects of actions and countermeasures), munition delivery errors and accuracy, damage mechanism and criteria, probability of kill, weapon reliability, and trajectory. The commander's intent and end state, desired effects, tasks, and guidance provide the basis for weaponeering assessment activities. Targeting personnel quantify the expected results of fires against prioritized targets to produce desired effects. Since time constraints may preclude calculations of potential effects against all targets, calculations should proceed in a prioritized fashion that mirrors the HPTL.

2-86. The weaponeering process is divided into several general steps and is not tied to a specific methodology or organization. The steps are not rigid and may be accomplished in different order or combined. The steps of the weaponeering process are—
- Identify collection requirements.
- Obtain information on friendly forces.
- Determine target elements to be analyzed.
- Determine damage criteria.
- Determine weapons effectiveness index.
- Determine aim points and impact points.
- Evaluate weapon effectiveness.
- Prepare preliminary documentation.
- Review collection requirements.

2-87. Collateral damage estimation is a methodology that assists the commander in staying within the law of war and rules of engagement. The law of war requires reasonable precautions to ensure only legitimate military objects are targeted. The law of war requires combatants to refrain from intentionally targeting civilian or noncombatant populations or facilities. The law of war also stipulates that anticipated civilian or noncombatant injury or loss of life and damage to civilian or noncombatant property incidental to attacks must not be excessive in relation to the expected military advantage to be gained. Failure to observe these obligations could result in disproportionate negative effects on civilians and noncombatants and be considered a law of war violation. Furthermore, United States leadership and the military could be subject to global criticism, which could adversely affect current and future military objectives and national goals.

2-88. During the targeting process, the staff has the responsibility to mitigate the unintended or incidental risk of damage or injury to the civilian populace and noncombatants, military personnel, structures in the immediate area, targets that are on the no-strike and/or restricted target list, livestock, the environment, civil air, and anything that could have a negative effect on military operations. This will assist the commander in weighing risk against military necessity and in assessing proportionality within the framework of the military decisionmaking process (MDMP).

2-89. Taking into account the weaponeering for a given target, the collateral damage estimation level 2 provides the assessment of whether a target meets the minimum requirement (criteria and approving authority) for employment of surface-to-surface scalable fires. A qualified individual with a current validation helps the commander and staff to evaluate collateral risk against targets during planning and the execution phases.

2-90. One method of mitigating collateral damage is reducing TLE by conducting target coordinate mensuration. Mensuration is the application of mathematical principles to a two dimensional surface in order to accurately determine the most accurate location of a target on all three planes of a Cartesian surface. Mensuration is applied to a target to reduce TLE. Correlating the expected target location to a highly refined coordinate reduces the TLE and provides a precise aimpoint that can be engaged with only the force necessary to achieve the desired effect.

2-91. The targeting working group must always determine the weapon system for targets of opportunity, subject to the maneuver commander's approval. All available attack assets should be considered. Attacking targets should optimize the capabilities of—
- Light and heavy ground forces.
- Attack aircrafts.
- Field artillery.
- Mortars.
- Naval gunfire.
- Combat air operations (both close air support and air interdiction).
- EW.
- Military information support operations.
- Civil affairs teams.

2-92. The availability and capabilities of each resource is considered using the following—
- Desired effects on the target.
- Payoff of the target.
- Degree of risk in the use of the asset against the target.
- Impact on friendly operations.

2-93. In some cases, the target attack must be coordinated among two or more weapon systems. Engagement of a target by indirect fires along with electronic attacks or monitoring may be of greater benefit than simply firing at the target.

TECHNICAL DECISIONS

2-94. Once the tactical decisions have been made, the fires cell directs the weapon system to attack the target. The fires cell provides the weapon system manager with the following—
- Selected time of attack.
- Effects desired in accordance with previous discussion.
- Any special restraints or requests for particular munitions types.

2-95. The weapon system managers—fires battalion S-3, air liaison officer, aviation, brigade naval gunfire liaison officer, and so on—determines if his system can meet the requirements. The fires cell is notified when a weapon system is unable to meet the requirements. There are various reasons a weapon system may not be able to meet the requirements, which may include the following—
- Systems not available at the specified time.

- Required munitions not available.
- Targets out of range.

2-96. The fires cell must decide if the selected weapon should attack under different criteria or if a different weapon should be used.

ASSESS

2-97. Commanders continuously assess the operational environment and the progress of operations, and compare them to their initial vision and intent. Commanders adjust operations based on their assessment to ensure objectives are met and the military end state is achieved.

2-98. The assessment process is continuous and directly tied to the commander's decisions throughout planning, preparation, and execution of operations. Staffs help the commander by monitoring the numerous aspects that can influence the outcome of operations and provide the commander timely information needed for decisions. The commander's critical information requirement is linked to the assessment process by the commander's need for timely information and recommendations to make decisions. Planning for the assessment process helps staffs by identifying key aspects of the operation that the commander is interested in closely monitoring and where the commander wants to make decisions.

2-99. Assessment occurs at all levels and across the spectrum of conflict. (See figure 2-8.) Even in operations that do not include combat, assessment of progress is just as important and can be more complex than traditional combat assessment. As a rule, the level at which a specific operation, task, or action is directed should be the level at which such activity is assessed. To do this, commanders and their staffs consider assessment ways, means, and measures during planning, preparation, and execution. This properly focuses assessment and collection at each level, reduces redundancy, and enhances the efficiency of the overall assessment process.

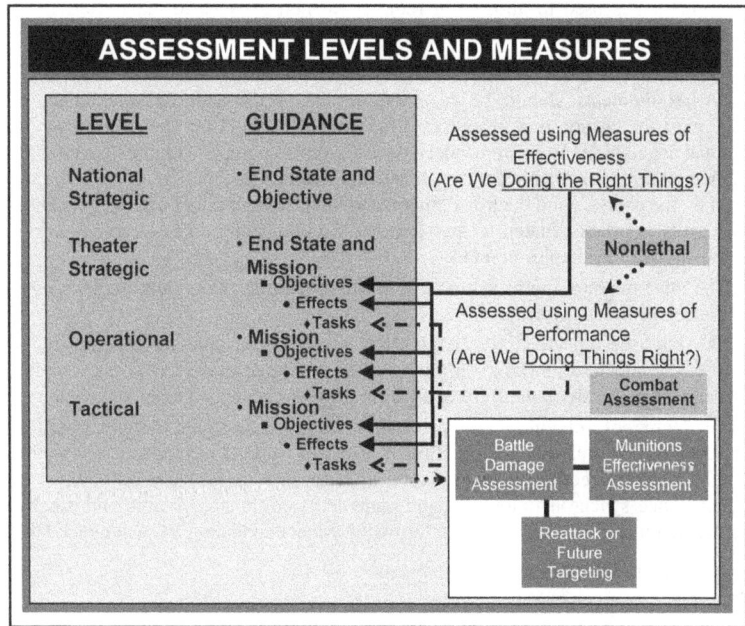

Figure 2-8. Assessment levels and measures

Chapter 2

COMBAT ASSESSMENT

2-100. Combat assessment is the determination of the effectiveness of force employment during military operations.

2-101. Combat assessment is composed of three elements—
- BDA.
- Munitions effectiveness assessment.
- Reattack recommendation.

2-102. In combination, BDA and munitions effectiveness assessment inform the commander of effects against targets and target sets. Based on this information, the threat's ability to make and sustain war and centers of gravity are continuously estimated. During the review of the effectiveness of air operations, redirect recommendations are proposed or executed.

Battle Damage Assessment

2-103. BDA includes known or estimated threat unit strengths, degraded or destroyed threat weapon systems, and all know captured, wounded, or killed threat personnel during the reporting period. BDA in the targeting process pertains to the results of attacks on targets designated by the commander. Producing BDA is primarily an intelligence responsibility, but requires coordination with operational elements to be effective. BDA requirements may be translated into PIR. BDA accomplishes the following purposes—
- At the tactical level, commanders use BDA to get a series of timely and accurate snapshots of their effect on the enemy. It provides commanders an estimate of the enemy's combat effectiveness, capabilities, and intentions. This helps the staff determine when, or if, their targeting effort is accomplishing their objectives.
- As part of the targeting process, BDA helps to determine if a restrike is necessary. The information is used to allocate or redirect weapon systems to make the best use of available combat power.

2-104. The need for BDA for specific HPT is determined during the decide function in the targeting process. BDA requirements should be recorded on the AGM and the intelligence collection plan. Commanders must be aware that resources used for BDA are the same resources used for target development and acquisition. The commander's decision must be made with the understanding that an asset used for BDA may not be available for target development and acquisition. BDA information is received and processed by the analysis and control element, and the results of attack are analyzed in terms of desired effects. The results are disseminated to the targeting working group. The targeting working group must keep the following BDA principles in mind—
- BDA must measure things that are important to commanders. Measurable things made available effortlessly are not priority.
- BDA must be objective. Receiving BDA product from another echelon, the G-2/S-2 should verify the conclusions (time permitting). The intelligence officers strive to identify and resolve discrepancies between the BDA analysts at different headquarters at all echelons.
- The degree of reliability and credibility of the assessment relies largely upon collection resources. The quantity and quality of collection assets influence whether the assessment is highly reliable (concrete, quantifiable, and precise) or has low reliability (best guess). The intelligence synchronization manager plans and coordinates organic and nonorganic collection assets to obtain the most reliable information when conducting BDA for each HPT.

2-105. Each BDA has three components. They are—
- Physical damage assessment.
- Functional damage assessment.
- Target system assessment.

2-106. These three assessments require different sensors, analytical elements, and timelines. They are not necessarily subcomponents of each BDA report.

Physical Damage Assessment

2-107. Physical damage assessment estimates the quantitative extent of physical damage through munitions blast, fragmentation, and/or fire damage effects to a target. This assessment is based on observed or interpreted damage.

Functional Damage Assessment

2-108. Functional damage assessment estimates the effect of attack on the target to perform its intended mission compared to the operational objective established against the target. This assessment is inferred based on all-source intelligence and includes an estimate of the time needed to replace the target function. A functional damage assessment is a temporary assessment (compared to target system assessment) used for specific missions.

Target System Assessment

2-109. Target system assessment is a broad assessment of the overall impact and effectiveness of all types of attack against an entire target systems capability; for example, enemy air defense artillery systems. It may also be applied against enemy unit combat effectiveness. A target system assessment may also look at subdivisions of the system compared to the commander's stated operational objectives. It is a relatively permanent assessment (compared to a functional damage assessment) that will be used for more than one mission.

2-110. BDA may take different forms besides the determination of the number of casualties or the amount of equipment destroyed. Other information of use to the targeting working group includes the following—

- Whether the targets are moving or hardening in response to the attack.
- Changes in deception efforts and techniques.
- Increased communication efforts as the result of jamming.
- Whether the damage resulting from an attack is affecting the enemy's combat effectiveness as expected.

2-111. Damage assessments may also be passive by compiling information in regards to a particular target or area. An example is the cessation of fires from an area. If BDA is to be made, the targeting working group must give intelligence acquisition systems adequate warning for sensor(s) to be directed at the target at the proper time.

2-112. BDA results may change plans and earlier decisions. The targeting working group must periodically update the decisions made during the decide function concerning the following—

- IPB products.
- HPTL.
- TSS.
- AGM.
- ISR plan.
- OPLAN.

Munitions Effectiveness Assessment

2-113. The G-3, in coordination with the fires cell and targeting working group, conducts munitions effectiveness assessment concurrently and interactively with BDA, as a function of combat assessment. Munitions effectiveness assessment is an assessment of the military force in terms of the weapon systems and munitions effectiveness. Munitions effectiveness assessment is conducted using approved weaponeering software and provides the basis for recommendations to increase the effectiveness of the following—

- Methodology.
- Tactics.
- Weapon system.
- Munition.
- Weapon delivery parameters.

Chapter 2

2-114. The G-3, in coordination with the fires cell, develops the munitions effectiveness assessment by determining the effectiveness of munitions, weapons systems, and tactics. Munitions effect on targets can be calculated by using approved weaponeering software. The targeting working group may generate modified commander's guidance to the assistant chief of staff, logistics/logistics staff officer concerning the following—
- Unit basic load.
- Required supply rate.
- Controlled supply rate.

Reattack Recommendation

2-115. Based on the BDA and munitions effectiveness assessment analysis, the G-2 and G-3 consider the level to which operational objectives have been achieved and make recommendations to the commander. Reattack and other recommendations should address operational objectives relative to the following—
- Target.
- Target critical elements.
- Target systems.
- Enemy combat force strengths.

ASSESSMENT METRICS AND MEASUREMENTS

Assessment Metrics

2-116. The staff should develop metrics to determine if operations are properly linked to the joint force commander's (JFC) overall strategy and the larger hierarchy of operational and national objectives. These metrics evaluate the results achieved during joint operations. Metrics can be objective (using sensors or personnel to directly observe damage inflicted) or subjective (using indirect means to ascertain results), depending on the metric applied to either the objective or task. Both qualitative and quantitative metrics should be used to avoid unsound or distorted results. Metrics can either be inductive (directly observing the operational environment and building situational awareness cumulatively) or deductive (extrapolated from what was previously known of the adversary and operational environment). Success is measured by indications that the effects achieved are influencing enemy, friendly, and/or neutral activity in desired ways among various target systems.

Measurement Types

2-117. The assessment process use measure of performance (MOP) and measure of effectiveness (MOE) to evaluate progress toward task accomplishment, effects creation, and objective achievement. Well devised measure can help the commanders and staffs understand the causal relationship between specific tasks and desired effects (JP 3-60).

The Targeting Methodology

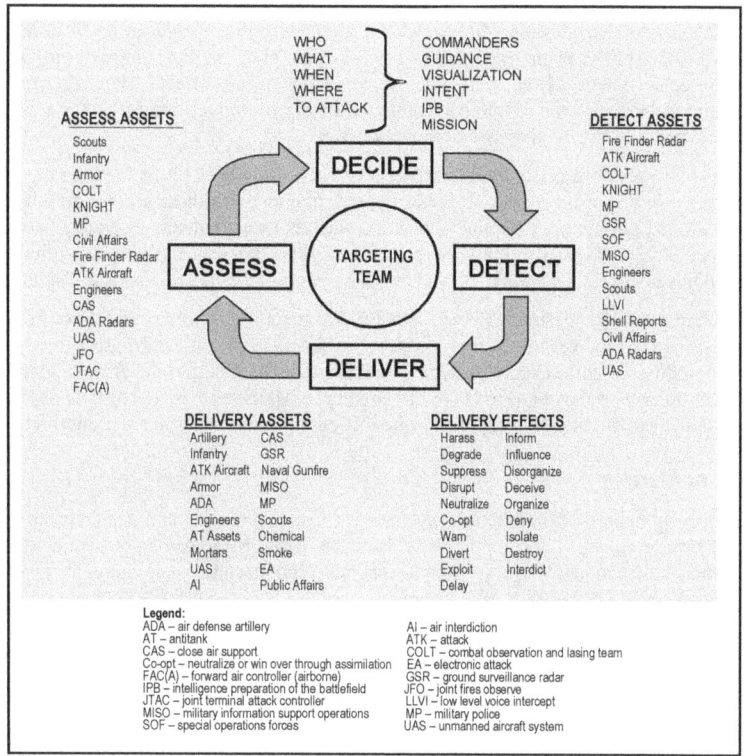

Figure 2-9. D3A methodology

Note. Figure 2-9 provides an example of assets available when conducting D3A methodology.

Measures of Performance

2-118. MOP answers the question "Are we doing things right?" and are the criteria for measuring task performance or accomplishment. MOP is quantitative, but can apply qualitative attributes to accomplish the task. Measurements are used in most aspects of combat assessment, since it typically seeks specific, quantitative data or a direct observation of an event to determine accomplishment of tactical tasks. MOP has relevance for noncombat operations as well as tons of relief supplies delivered or noncombatants evacuated. MOP is used to measure operational and strategic tasks, but the type of measurement may not be as precise or as easy to observe (JP 3-60).

2-119. MOP helps answer questions like, "was the action taken, were the tasks completed to standard, or how much effort was involved?" Regardless of whether there was or was not a tactical, immediate effect, "did the assigned force execute the 'fires,' 'maneuver,' or 'information' actions as required by the specified or implied task?" MOP is used by the commander to assess whether his directives were executed by subordinate units as intended or if the units were capable of completing the specified action. Typical measures might include the following—

- Did the designated unit deliver the correct ordinance?
- How many leaflets were dropped, what type of leaflets, and did they saturate the target?
- How many potholes were filled and include the time taken to complete the task?
- How much potable water was delivered to the village?

Measures of Effectiveness

2-120. MOE answers the question, "are we doing the right things?" and are used to assess changes in system behavior, capability, or the operational environment. They are tied to measuring the attainment of an end state, achievement of an objective, or creation of an effect. They do not measure task accomplishment or performance. While MOE may be harder to derive than MOP for a discrete task, they are nonetheless essential to effective assessment (JP 3-60).

2-121. MOE indicates progress toward attainment of each desired effect or indicate the avoidance of an undesired effect. MOE is a direct form of measurement, like an eyewitness account of a bridge span being down; some may be more circumstantial indicators, such as measurements of traffic backed up behind a downed bridge. MOE is typically more subjective than MOP but can be crafted as either qualitative or quantitative indicators to reflect a trend as well as show progress relative to a measurable threshold.

2-122. For example, if the desired effect is that Brown government forces withdraw from the cities, the MOE could be stated as increase or decrease in level of forces in the cities. Progress toward this effect can be measured intelligence collection. However, if the desired effect is that the Brown government engages the terrorists in order to cause them to leave the country, a MOE such as increase or decrease in coercive content of diplomatic communiqués could be more difficult to track, measure, and interpret.

Characteristics of Metrics

2-123. Assessment metrics should be relevant, measurable, responsive, and resourced so there is no false impression of task or objective accomplishment. Both MOP and MOE can be quantitative or qualitative in nature, but meaningful quantitative measures are preferred because they may be less susceptible to subjective interpretation (JP 3-60).

Relevant

2-124. MOP and MOE should be relevant to the task, effect, operation, the operational environment, the desired end state, and the commander's decisions. This criterion helps avoid collecting and analyzing information that is of no value to a specific operation. It also helps ensure efficiency by eliminating redundant efforts.

Measurable

2-125. Assessment measures should have qualitative or quantitative standards they can be measured against. To measure change effectively, a baseline measurement should be established prior to execution to facilitate accurate assessment throughout the operation.

Responsive

2-126. Assessment processes should detect situation changes quickly enough to enable effective response by the staff and timely decisions by the commander. Assessors should consider the time required for an action or actions to take effect within the operational environment and for indicators to develop. Many actions require time to implement and may take even longer to produce a measurable result.

Resourced

2-127. To be effective, the assessment process must be adequately resourced. Staffs should ensure resource requirements for collection efforts and analysis are built into plans and monitored. An effective assessment process can help avoid duplication of tasks and avoid taking unnecessary actions, which in turn can help preserve military power.

Chapter 3
Corps and Division Targeting

Targeting at corps and division level may be at the tactical or operational levels of war. It involves commanders and staff officers in the decide, detect, deliver, and assess (D3A) methodology in support of tactical operations.

The design of the Army corps headquarters provides a core on which to build a combined or joint force land component headquarters. The corps headquarters becomes the Army forces for all Army units including those units not under operational control of the corps. The corps has operational control of the Army division and tactical control of the Marine Corps division. The composition of the headquarters should roughly reflect the composition of the joint and multinational land forces involved. While acting as a joint forces land component command (JFLCC) headquarters, the corps headquarters will also perform duties of the Army forces headquarters and normally have operational control of subordinate Army units. Like the theater army operational command post (CP), the corps will use Army doctrine and procedures but refer to joint doctrine. The corps headquarters constituted and so designated as a joint task force (JTF) by the Secretary of Defense or a combatant commander. Joint interdependence requires Army leaders and staff to understand joint doctrine and procedures when participating in joint operational planning and assessment.

Normally, the division will be under the operational control of a JFLCC, the Army forces, or a corps headquarters for major combat operations. For small-scale operations, it may be a JFLCC or the Army forces operations control of a JTF.

FIRES CELL

3-1. The primary action agency for targeting at the corps and division level is the fires cell. The fires cell coordinates available weapon systems that provide Army indirect fires, joint fires, electronic attacks, and the associated targeting process. The fires cell implements the commander's intent through the physical attacks on enemy capabilities or the degradation of enemy command nodes and control systems. An example of a fires cell is shown in figure 3-1.

3-2. The fires cell uses the D3A methodology and understands the joint targeting cycle. At a minimum, the fires cell will be responsible for nominating targets for inclusion in the joint targeting process and must understand the targeting deadlines and how to influence the process to achieve the corps or division commander's objectives.

3-3. The fires cell synchronizes all shaping fires and directs the attack of targets by organic or attached fire support. This includes synchronizing fires for joint suppression of enemy air defense (J-SEAD) to support air and aviation operations. It coordinates the use of airspace through the airspace command and control element. It coordinates air support requirements through the air support operations center and corps or division tactical air control party (TACP).The fires cell coordinates directly with the battlefield coordination detachment.

3-4. There are two fires cells in the corps and division, one cell at the main CP and another at the tactical CP. The fires cell at the main CP has four elements—
- Fires element.
- Current operations cell.

Chapter 3

- Electronic attack section.
- Field artillery intelligence officer.

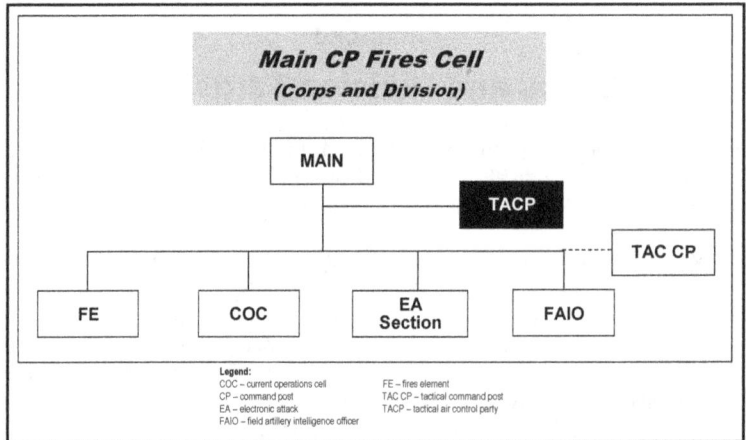

Figure 3-1. Fires cell

FIRES ELEMENT

3-5. The fires element synchronizes joint, interagency, multi-national assets, fire support, and sensor management. It provides input to the intelligence collection plan, intelligence, surveillance, and reconnaissance (ISR) synchronization matrix, and the targeting process.

3-6. Specific functions include—
- Synchronize joint, interagency, and multinational assets.
- Recommend distribution decision of close air support assets to commander.
- Provide access to joint fires to interagency and multi-national forces.
- Prioritize and allocate fires resources.
- Conduct airspace coordination.
- Conduct fire support knowledge management.
- Provide input into the joint air tasking cycle that produces the air tasking order/airspace control order/special instructions.
- Coordinate with other Service components.
- Maintain fires common operational picture.
- Time-sensitive target (TST) nomination, management, and execution.
- Conduct target development.
- Conduct target coordinate mensuration when applicable.
- Conduct munitions effects analysis (weaponeering) when applicable.
- Conduct collateral damage estimation when applicable.
- Conduct target management.
- Review and comply with rules of engagement.
- Maintain list of high-value individuals (HVI).
- Develop and maintain the high-payoff target list (HPTL).
- Monitor and nominate targets to the restricted target list and no-strike list.
- Consolidate, prioritize, and nominate targets for inclusion in the joint integrated prioritized target list.
- Develop target selection standards (TSS).

- Provide input to collection plan; synchronizing ISR assets with designated targets.
- Conduct combat assessments.
- Integrate and synchronize fire support and synchronize cyber-electromagnetic activities into D3A process.
- Advise on application of joint fires.
- Chair candidate target list review board.

3-7. If the corps or division headquarters becomes the Army forces, JTF, or JFLCC headquarters, the fires element may also be required to—
- Serves as the core for a JTF joint fires element.
- Conduct sensor management and synchronization.
- Interface with battlefield coordination detachment and higher joint fires element.
- Identify fires effects requirements from other components (air interdiction/naval surface fires).
- Review and comment on the joint force air component commander's (JFACC) apportionment recommendation.
- Recommend JFLCC assets for the joint force commander's (JFC) allocation (Army Tactical Missile System/attack helicopter).
- Advise on fires asset distribution (priority) to land forces.
- Develop JFLCC priorities, timing, and effects for air interdiction within the joint operational area.
- Develop JFLCC targeting guidance and priorities.
- Develop the JFLCC command target lists and fire support coordination measures (FSCM).
- Execute the joint targeting coordination board (JTCB) and the JTCB working group.

CURRENT OPERATIONS CELL

3-8. This cell executes current operations, prepares, and sets conditions necessary for future operations simultaneously.

FIELD ARTILLERY INTELLIGENCE OFFICER ELEMENT

3-9. As a participant in the division, corps, and joint targeting process, the field artillery intelligence officer (FAIO) coordinates with corps internal and external all-source intelligence elements to provide input to the development, nomination, and prioritization of targets.

3-10. Specific functions include—
- Develop and nominate priority targets.
- Participate in the joint targeting process and cycle.
- Ensure that targets are prioritized and sequenced in current operations and future plans.
- Coordinate with the analysis and control element for all source target information to develop target nominations.
- Operate computers systems used in the targeting process.
- Develop the sensitive target approval and review packets.

ELECTRONIC ATTACK SECTION

3-11. The electronic attack section serves as the principal planning section on the corps or division staff. The section plans, coordinates, integrates, deconflicts, and assesses the use of physical attack, electronic attack, EW support, computer network attack, and computer network exploitation for current and future operations. These capabilities are intended to degrade, destroy, and exploit an adversary's ability to use the electromagnetic spectrum, computers, and telecommunication networks.

Chapter 3

3-12. Specific functions include—
- Plan, coordinate, integrate, synchronize, and deconflict electronic attack, electronic protection, electronic support, and physical attack.
- Be familiar with the EW support to current operation plan (OPLAN) and concept plans.
- Recommend and promulgate electronic attack special instructions and rules of engagement.
- Archive electronic attack planning, execution data, and lessons learned.
- Identify and coordinate intelligence support requirements for joint electronic attack activities.
- Coordinate with ISR assets and national agencies in assessing the adversary's ability to use electromagnetic spectrum, computers, and telecommunication networks.
- Plan, coordinate, and assess electronic support requirements.
- Plan, coordinate, and assess electronic deception.
- Coordinate spectrum management and radio frequency deconfliction related to joint, interagency, intergovernmental, and multinational EW.
- Plan, assess, and coordinate friendly electronic security measures related to electronic attack activities.
- Plan, request, employ, and synchronize nonorganic joint, interagency, intergovernmental, and multinational EW capabilities.
- Employ joint capabilities at the tactical level for shared awareness and unhindered collaboration.
- Maintain current assessment of EW resources available to the JFC.
- Plan and prioritize scalable fires and electronic attack.
- Predict effects of friendly and enemy EW operations.
- Integrate electronic attack and intelligence preparation of the battlefield (IPB) into the military decisionmaking process (MDMP).
- Serve as the jamming control authority.
- Implement EW policies to control the electromagnetic spectrum or to attack the enemy.

3-13. If the corps becomes an Army forces, JTF, or JFLCC headquarters, the EW coordination cell may also be required to—
- Serve as the JFC EW representative.
- Serve as the principal EW planner on the joint staff as part of the operational directorate of a joint staff (J-3) and coordinate with the command's information operations cell.
- Serve as a member of the standing joint planning group for EW planning and execution.
- Serve as joint EW coordination center to plan operational level EW for the JFC.
- Coordinate and monitor joint coordination EW reprogramming.
- Prepare EW portion of estimates and tabs to joint force OPLAN.
- Participate in joint targeting board to formulate and recommend EW targets.
- Participate in joint boards/cells.

TACTICAL COMMAND POST FIRES CELL

3-14. Coordinated fires at the corps tactical CP. Specific functions include—
- Execute fires plan in support of a specific operation.
- Request and coordinate close air support and air interdiction.
- Synchronize scalable fires.
- Conduct lethal fires, assess, and reattack recommendations.
- Coordinate with maneuver and control element.
- Synchronize joint, interagency, multinational assets.
- Interface with battlefield coordination detachment and higher joint fires element.
- Coordinate with other components.
- Coordinate EW activities.

REQUIREMENTS FOR SUCCESSFUL TARGETING

3-15. The operating environment and targeting capabilities influence the D3A methodology. Planning is different for a conventional war against a sophisticated enemy, requiring interdiction of operational targets, than that for stability operations against an insurgent force where targets are difficult to locate. With evolving security threats, each corps and division staff is concerned with several contingency plans. High-value targets (HVT) and HPTs are developed for plans that are regional and for which adequate intelligence is available. In addition, for planning purposes, each contingency has an associated list of forces that contains listings of available nonorganic collection and delivery assets.

3-16. Targeting is the process of selecting targets and matching the appropriate response to them—scalable capabilities—based on the mission, enemy, terrain and weather, troops and support available, time available and civil considerations (METT-TC). The targeting process is an integral part of the way Army headquarters solve problems. Both the joint targeting cycle and the D3A methodology facilitate the MDMP. The D3A methodology is merely a mechanism for grouping the targeting tasks which must occur.

3-17. Targeting is done throughout the current and anticipated areas of interest. The operational success of the corps and/or division battle depends on—
- The commander's battle plan.
- The timeliness and accuracy of intelligence from national, theater, corps, and division assets.
- The speed with which the corps or division achieves and exploits its tactical and operational advantages.
- The ability of the staff to synchronize a multi-Service targeting effort.

3-18. The corps targeting working group is a planner and an executor of the targeting process. It has the assets needed to see, plan, and execute targeting for shaping operations while synchronizing targeting in support of decisive and sustainment operations. To engage the enemy, it involves the coordinated use of all of the following—
- Intelligence.
- Surface-to-surface fires.
- Army aviation.
- Air component.
- Special operations forces.
- Unmanned aircraft system.
- Navy and Marine Corps assets.
- Joint terminal attack controller and/or joint fires observer.

3-19. The corps uses the collection assets in the surveillance brigade to collect data throughout the area of operations. The corps targeting working group also has various systems that link it to echelons above corps and national collection and weapon systems. The corps main CP has communications, computers, and intelligence elements to synchronize the overall operations and long-range targeting. The below elements are critical to the targeting process—
- Field artillery units.
- Army attack reconnaissance aviation.
- Special operations forces.
- Electromagnetic activities.
- Air component.
- Naval assets.
- Joint terminal attack controller and/or joint fires observer.

3-20. With these links, the fires cell can aggressively attack the commander's HPT.

Chapter 3

D3A METHODOLOGY

3-21. The actions and functions of the corps and division targeting working groups are essentially the same with the chief difference being the capabilities of the assets available for targeting. The division relies heavily upon corps and echelons above corps assets for targeting support for its shaping operations.

3-22. The commander directs the targeting effort. The process begins with the commander's guidance after the G-3 and G-2 present their initial mission analyses. Along with his mission statement, the commander must give his guidance on—
- What he expects the unit to do.
- What he feels are the most important targets.
- What general effects he wants to have on those targets.

3-23. One individual must supervise the targeting process. In the main CP, the chief of fires is responsible for supervising the targeting process and the targeting working group. At corps level, the G-3 could be an alternative supervisor. The targeting working group incorporates the mission statement the commander's intent, and the concept of the operation into the target value analysis process.

3-24. Once the staff has this information, the targeting working group analyzes enemy course of action (COA) and identifies basic HVT at the same time. As the staff war games friendly COA, the targeting working group develops initial proposals on HPT and attack guidance. After the commander selects the final COA and issues further guidance, the targeting working group—
- Refines and prioritizes the HPTL.
- Develops the attack guidance matrix (AGM).
- Submits these products to the commander for approval.

3-25. Once approved, the HPTL and AGM (form the basis for the activities of the targeting working group. The G-3 ensures that the intelligence, operations, plans, and fires cells incorporate these products into the operation order (OPORD) and its annexes. For example, included are tasks to subordinate units, coordinating instructions, and priority intelligence requirements (PIR). The G-2, G-3, and chief of fires determine what additional support is required for collection and target attack. When support requirements have been determined, they submit the appropriate requests.

3-26. The targeting working group provides TSS to the fires cell by using the following—
- Timeliness.
- Target status (stationary or moving).
- Target characteristics consist of size, accuracy, and target location error (TLE) requirements for each weapon systems.

3-27. The targeting working group also determines the targets that require BDA. Only the most critical targets should be selected, as valuable assets must be diverted from target or situation development to perform BDA.

3-28. The G-2 ensures appropriate HPT is approved as PIR and a collection plan that focuses on answering the commander's PIR is developed. The collection management section provides targeting information to the intelligence analyst for analysis. The FAIO helps the analyst in this process. FAIO provides knowledge of requirements for identifying the most important and perishable targets. The FAIO and analyst inform the targeting working group when major changes in the tactical situation warrant reevaluation of the HPTL. The targeting working group continually assesses the current situation and future needs. At the same time, the team reevaluates the HPTL, AGM, BDA requirements, and TSS and updates them as necessary. The FAIO works closely with the collection management section as well. The FAIO helps that section translate targeting working group requirements into guidance for the collection plan and provides expertise on field artillery target acquisition systems.

3-29. The analysis and control element and FAIO evaluate the information from the collection management section against the TSS and HPTL to determine targets or suspected targets. Targets are immediately passed to the fires cell for attack. Enemy activities that do not achieve TSS are suspected targets. Enemy activities that appear on the HPTL but categorized as suspected targets are passed to the fires cell for correlation with information available at the fires cell. This correlation may produce a valid target. Also, the FAIO should request the collection manager focus additional collection assets to further develop selected suspected targets. The position coordinates with the collection manager to retrieve BDA data as acquired.

3-30. Nonlethal targeting is a critical part of the targeting methodology. Staff members responsible for military deception, public affairs, EW activities, information protection, and operations security attend the targeting working group sessions. These capabilities are coordinate, integrate, and synchronize into the nonlethal targeting effort in support of tactical operations. The same process used to determine when a radar system should be attacked with the Army Tactical Missile System is also used to determine when building a new sewer system will influence local leaders to support friendly objectives. All available weapons system must be completely integrated to ensure every effort is directed toward achieving the commander's desired effect.

3-31. The EW officer integrates the electromagnetic activities as part of the overall military operations. The electromagnetic activities are an integral part of the targeting process. The electromagnetic activities effort divided the electronic warfare into three actions: electronic attack, electronic protection, and electronic support. The actions of electronic warfare are to seize, retain, and exploit an advantage over adversaries and enemies across the corps and division electromagnetic spectrum. These actions include also denying and degrading adversary and enemy information operations and protecting friendly mission command networks and systems. The EW officer and staff coordinate their efforts with the targeting working group to accomplish the commander's objectives. The EW officer provides recommendations before the technical control element receives the collection plan and the division EW composite target list. The limited allocation of intelligence and EW assets causes conflict between the collection plan and the division EW composite target list. The G-3 finalizes any conflicting recommendations between G-2, EW officer, and the technical control element.

3-32. The fires cell receives most target nominations from the FAIO. Once a target is received, the fires cell analyzes it in terms of TSS and the AGM, prioritizes it, and determines an appropriate attack method. The fires cell may consult with other agencies to facilitate target engagement. This is especially necessary when weapons system availability, rules of engagement, or other considerations determine the method of attack. Coordinated attacks or any combination of scalable fires may necessitate temporary augmentation of the fires cell. The fires cell directs the selected attack unit to engage the target and provide BDA data through the G-3 or representative of the unit at the main CP. The all-source analysis section and FAIO analyze BDA data for selected targets to evaluate the effectiveness of the attack. However, the targeting working group determines whether the commander's attack guidance has been achieved or further fires are necessary.

3-33. Targets and missions beyond the capability of the corps or division to properly service with their assets are passed to higher headquarters for action. The staff must know when the requests must be submitted for consideration within the requested echelon target planning cycle. The synchronization of these missions with ongoing operations may be critical to the success of the unit mission. Close coordination between supported and supporting components is required to ensure vertical integration and synchronization of plans. A key to coordination for both planning and execution is the use of liaison officers at all headquarters.

3-34. During this process, the commander, chief of staff, G-2, G-3, Army aviation commander, and chief of fires exert considerable influence. Targeting is a process that involves the entire staff. Leaders must keep the targeting effort focused so that the targeting cells devote their fullest efforts to the process.

3-35. The targeting process is a continuous and cyclical effort. Phases occur at the same time when executing current operations and planning future operations. The phases are sequential in the context of any given planning cycle. Recurring events and their associated products are best managed through workable standard/standing operating procedures (SOP). SOPs must be tailored to the unit's structure and operating environment to ensure a cohesive, coordinated targeting effort. A sample SOP for a targeting working group at corps or division level is located in Appendix F.

CORPS AND DIVISION SYNCHRONIZATION

3-36. The focus is on shaping targets and operational targets at the corps level. These targets must be engaged to shape the battlefield for the decisive battle at division level and below. The corps long-range assets must be integrated and synchronized with joint force systems. Simultaneously, the corps commander must support detect and deliver requirements of subordinate units. Corps, and perhaps division, HVT located throughout the depth of the battlefield. Corps and division commanders set the targeting priorities, timing, and effects consistent with the higher commander's guidance. Mission analysis and plan development establish what conditions must be achieved for success. The mission analysis determines the combat activities, sequence of activities, and application of resources that will achieve the conditions for success. While all conditions may not be met, the commander is responsible for the coordination and synchronization of supporting Service and joint assets in his area of responsibility.

3-37. The corps ensures subordinate divisions and separate units understand the corps mission and concept of operations. Each division plan supports the corps commander's intent and guidance. The corps shaping operations establish the conditions for the divisions to fight the corps commander's decisive operations. This understanding between corps and division means that each command supports the other. Missions and targets may be passed from corps to divisions as the more appropriate executor. The divisions may also have missions and targets that are beyond their capabilities that require the corps to provide support. This is important considering the limited range of division assets to detect and attack targets. The corps may coordinate attack of corps HPT in a division area; similarly, the division may ask the corps to acquire division HPT beyond the capability of the division. This mutual support must be coordinated and synchronized during the decide phase of the planning process. Synchronization includes all of the following—

- Coordinating the acquisition, tracking and reporting of targets of concern at either, or both echelons.
- Submit airspace control means requests to coordinate airspace requirements.
- Vertical exchange of target information.
- Attack of targets outside the area of operations of an echelon.
- Target engagement criteria.
- Allocation of assets.
- Establishment of communications links between sensor systems, decisionmakers, and weapon systems.

Note. An example is the acquisition of a corps HPT by division assets that is reported to corps and attacked by corps assets.

3-38. The fires cell, subordinate units, and Service components liaisons play key roles in the synchronization process.

AIR-GROUND OPERATIONS AT CORPS LEVEL

INTERDICTION

3-39. Interdiction is an action to divert, disrupt, delay, or destroy the enemy's military potential before it can be used effectively against friendly forces or to otherwise achieve objectives (JP 3-03). A successful interdiction requires a combination of other Service components' resources and a robust liaison between these components. The Service components combined fundamental actions to divert, disrupt, delay, or destroy the enemy should be conducted across the full breadth and depth of the area of operations. The Service component commanders' staff creates challenging and simultaneous demands on the enemy and their resources. All of these actions and collective efforts create, enhance, and sustain the friendly scheme of maneuver for a joint interdiction.

3-40. The commander completes the mission analysis that includes guidance, intent, and desired effect. The fires cell elements conduct a preliminary analysis of the mission to identify factors pertaining to ground forces. The mission is restated and planning guidance is given to the targeting element for their consideration when preparing their running estimates. The targeting element identifies the high-value targets for scalable fires and actions as in air interdiction, suppression of enemy air defense (SEAD), special operations forces, and strategic attack to contribute to a successful outcome of a joint campaign or major operations.

3-41. The fires cell interfaces with the battlefield coordination detachment assigned to the appropriate joint air operations center or a combined air operations center. Tasks include exchanging current intelligence and operational data, support requirements, and coordinating Army forces' requirements for airspace control measures, FSCM, and air support operations requests. The battlefield coordination detachment serves as the Army forces' liaison to articulate the commander's request for air operations support for ground operations to complement the JFC end state. The main CP fires cell is the targeting process link to planning, coordinating, integrating, and synchronizing ground forces into the joint integration operations.

THEATER AIR-GROUND SYSTEM

3-42. The theater air-ground system provides liaison elements from corps down to battalion level. See FM 3-52.2 for complete details.

Tactical Air Control Party

3-43. The TACP at corps and lower levels provide advice and planning assistance on the employment of air support. The TACP works with the fire support elements at each level. At corps and division level, the TACP consists of the following liaison personnel—

- Air liaison officers.
- Joint terminal attack controllers.
- Airlift mobility liaison officers.
- Reconnaissance liaison officers.

3-44. The Air Force air request net is used by TACP at all levels and the air support operations center to request and coordinate close air support operations. When the COMAFFOR is also the JFACC, the JFACC will augment the air operations center with elements from other components to create a joint air operations center. When the air operations center becomes a joint air operations center, the Air Forces air request net becomes the joint air request net.

3-45. The TACP is supervised by the air liaison officer and performs the following functions—

- Serves as the Air Force commander's representative, providing advice to the commander on the capabilities, limitations, and employment of air support, airlift, and air reconnaissance.
- Provide a coordination interface with the respective fires cell and airspace command and control element. Help synchronize air and surface fires and helps prepare the air support plan. Provide direct liaison for local air defense and airspace management activities.
- Advise, help develop, and evaluate close air support, interdiction, reconnaissance, and SEAD targets.

Air Support Operations Center

3-46. The air support operations center is directly subordinate to the air operations center and is responsible for the coordination and control of air component missions in its assigned area requiring integration with other supporting arms and ground forces. The air support operations center processes immediate close air support requests received over the joint air request net, coordinates execution of preplanned, and immediate close air support and normally exercises tactical control of joint forces made available for tasking. Once the ground element approves immediate requests, the air support operations center tasks on-call missions or diverts—with ground element approval—scheduled missions. The air support operations center may be granted launch and/or divert authority over all or some of these missions. If the air support operations center has not been given control of on-call or scheduled missions, they must contact the air operations center or joint air operations center to launch or divert close air support missions. (See JP 3-09.3 for more information on the air support operations center.)

3-47. The air support operations center is responsible for establishing and maintaining the theater air control system at levels below the corps.

Joint Air Operations Center

3-48. The personnel assigned to the joint air operations center normally pass control of close air support missions to the air support operations center. The air support operations center passes requests that cannot be satisfied with previously distributed assets to the joint air operations center. The JFACC normally retains control over all air interdiction and reconnaissance operations supporting the corps as directed by the JFC. The JFACC has the authority to redirect aircraft to support immediate air support requests.

Airspace Command and Control

3-49. The airspace command and control element is the Army's operational approach to accomplishing the functional activity of airspace control. The airspace command and control element enhances the synchronization of forces using airspace. This element is part of the commander's staff and participates in the Army and joint tasks associated with the mission command warfighting function at each echelon of the Army. The team effort is required to coordinate and integrate airspace user requirements during plans and operations. Graphic control measures include airspace coordinating measures that regulate airspace users and are integrated with other graphic control measures and FSCM. The airspace requirements are the commander's responsibility and extend down through all tactical command levels to the maneuver brigade. The airspace element integrates key staff, missile defense, fires cell, and air traffic service, to include the air maintenance unit and liaison personnel. The Army air-ground system works in conjunction with the theater air control system to coordinate and integrate both Army component aviation support and Air Force component support with Army ground maneuver.

3-50. The airspace command and control element is under the direction of the G-3 air, and is the primary lead in the planning and management of airspace over the ground battle at corps and division. The airspace element staffer integrates and helps synchronize all functional operations that share airspace with other friendly forces, including the following—

- Fire support.
- Air and missile defense.
- Army aviation.
- Intelligence.
- Intra-theater airlift.
- Amphibious.
- Joint and multinational.

TARGETING RESPONSIBILITIES

3-51. The Army targeting responsibility begins with the commander and the process include the G-3/S-3, fire support, airspace, intelligence, and other supporting staff and liaison personnel. The formal structure of the elements at corps and division depend on the operating environment. Tailoring the formal structure of the staff working environment is necessary to ensure a cohesive, coordinated targeting effort. Key personnel and their targeting responsibilities are listed below.

COMMANDER

3-52. The commander issues guidance on the concept of operation. The concept of operations and mission defined by the commander's intent facilitates a shared understanding and focus for the targeting working groups. The targeting working group relies on individual initiative and coordination to act within the concept of operations.

CHIEF OF STAFF

3-53. The chief of staff is responsible for supervising the targeting process and chairs targeting boards.

CHIEF OF FIRES

3-54. The chief of fires is a functional staff officer serving at division and corps level who is responsible for advising the commander on the best use of available fire support resources, providing input to necessary orders, and developing and implementing the fire support plan. Normally, the chief of fires is the colonel or senior lieutenant colonel in charge of the fires cell.

3-55. The chief of fires is essential to effective fire support. The position is responsible for the operations and training of the fires cell as well as contributing to the work of the current operations, future operations, and plans cell as required. The corps fires cell coordinates available weapon systems that provide Army indirect fires, joint fires, EW activities, and the associated targeting process. Specific duties at the modular corps headquarters include—

- Plan, coordinate, and synchronize all aspects of fire support—
  - Physical attack/strike operations.
  - Electronic attack.
  - Electronic protection.
  - Electronic support.
- Advise the corps commander and staff of available fire support, including capabilities and limitations.
- Chair the targeting working group.
- Participate in the MDMP.
- Work with the commander, deputy commander, and chief of staff to integrate scalable indirect fires and joint fires into the concept of operations.
- Develop, recommending, and briefing the concept of fires to the corps commander, and preparing the fires paragraph of all OPLAN/OPORD.
- Coordinate training of force subordinate organization fires cells with their respective maneuver units and with the fires battalions.
- Accompany the force commander in the command group during execution of tactical operations.

DEPUTY CHIEF OF FIRES

3-56. The deputy chief of fires provides the latest status of fire support resources and finalizes the attack guidance formulated by the commander and the chief of fires. The deputy chief of fires specific actions include the following—

- Coordinate the functions of the targeting working group.
- Recommend target priorities for acquisition and attack based on target value analysis and war gaming.
- Recommend to the chief of staff methods of attack for targets.
- Support by the other members of the targeting working group, develop the HPTL, AGM, and BDA requirements.
- Develop timeliness and accuracy guidelines for the TSS for use by the FAIO and the fires cell with the G-2 plans/operations officer.
- Assist by the EW officer to develop targets for electronic attack.
- Monitors changes in the situation and reassess the HPTL, AGM, timeliness, and accuracy guidelines of the TSS, and BDA requirements.
- Synchronize timing of attack with the G-3 and subordinate units.

Chapter 3

- Coordinate support for subordinate units attack requirements.
- Coordinate SEAD, J-SEAD, and joint air attack team.
- Receive BDA and, with the G-2 and G-3, determines if an attack resulted in the desired effects or if additional attacks are required.
- Ensure target nominations meet validation review for integration on the joint integrated prioritized target list.

INTELLIGENCE OFFICER

3-57. The G-2 synchronizes the ISR plan and provides information on the current enemy situation as well as provides estimates as to what the enemy is capable of doing in the future. The position provides assessments of probable enemy actions, analyzes, and identifies targets based on the commander's guidance. The G-2 more specific actions are as follows—

- Develop and monitors the enemy situation.
- Develop and provides IPB products to the other targeting working group members.
- Pass HPT and suspected HPT to the fires cell.
- Develop HVT.
- Develop the HPTL, AGM, and BDA requirements with the other members of the targeting working group.
- Distribute the intelligence collection plan to collection managers.
- Provide input to the fires cell on TSS.
- Periodically reassess the HPTL, AGM, and BDA requirements with the deputy chief of fires and G-3 plans/operations officer.
- Receive BDA and, with the deputy chief of fires, determines if an attack resulted in desired effects or if additional attacks are required.
- Provide input for the decision support template.
- Participate in the extensive intra-staff coordination with the rules of engagement.

OPERATIONS OFFICER

3-58. The G-3 officer's actions are as follows—

- Develop the HPTL, AGM, and BDA requirements and ensures they are integrated with the decision support template with the other members of the targeting working group.
- Concentrate on future and contingency operations.
- Ensure the plans reflect the commander's concept of operation.
- Periodically reassess the HPTL, AGM, and BDA requirements with the deputy chief of fires and G-2 plans/operations officer.
- Determine if an attack resulted in the desired effects or if additional attacks are required with the deputy chief of fires and G-2 plans/operations officer.
- Coordinate the implementation of tailored rules of engagement to support national policies.

OPERATIONS AIR

3-59. The G-3 air's actions are as follows—

- Supervise the airspace element.
- Approve the airspace annex.
- Prioritize and integrate airspace users and airspace control means requests.
- Synchronize airspace requirements with warfighting functions during the operations process.
- Coordinate the integration of tactical airlift.

SPECIAL OPERATIONS LIAISON

3-60. The Army special operations forces commanders may establish or receive liaison and coordination elements with higher and adjacent units or other agencies, as appropriate. These elements may include a special operations command and control element, the special forces liaison element, or a special operations liaison element. The element may consist of a single Soldier. When provided, the special operations forces liaison actions are as follows—
- Forward target nominations and missions requirements to the JFC for consideration by the JTCB.
- Coordinate JTCB tasking with the joint force special operations component commander for feasibility assessment and execution.

FIELD ARTILLERY INTELLIGENCE OFFICER

3-61. The FAIO actions include the following—
- Collocate with the G-2 staff particularly the collection manager and all-source analysis section.
- Expedite targeting information from the analysis and control element to the fires cell.
- Monitor the enemy situation and keeps the deputy chief of fires informed. Recommends changes to priorities and attack means.
- Provide input concerning the threat, TSS, attack guidance, and list of HPT types.
- Supervise or conduct target coordinate mensuration when applicable.
- Supervise or conduct weaponeering when applicable.
- Supervise or conduct collateral damage estimation when applicable.
- Provide information to the intelligence cell regarding accuracy requirements and timeliness of information for the fire support system.
- Ensure essential target information is compared to TSS prior to passing a target to the fires cell.
- Advise the deputy chief of fires when changes in the situation warrant reassessment of the HPTL and AGM.

CORPS AND DIVISION TARGETING OFFICER

3-62. The corps and division targeting officer's actions follow—
- Deploy to the corps or division main CP to help form the fires cell.
- Advise and keeps the chief of fires informed on issues concerning targeting and fire support.
- Participate as a member of the targeting working group at corps or division.
- Help determine the HPTL.
- Help determine the AGM.
- Help determine the TSS.
- Supervise or conduct target coordinate mensuration when applicable.
- Conduct munitions effects analysis (weaponeering) when applicable.
- Supervise or conduct collateral damage estimation when applicable.
- Interface with the fires cell in subordinate units.
- Keep the FAIO informed on changes to the HPTL, TSS, and AGM.
- Pass targets received from the FAIO to weapon systems in the most expedient manner.
- Perform duties as a targeting officer at the battlefield coordination detachment when required.

INTELLIGENCE PLANS/OPERATIONS OFFICER

3-63. The G-2 plans/operations officer's actions follow—
- Maintain a current enemy situation map.
- Maintain the target database.
- Evaluate and analyzes combat information with the FAIO to identify HVT and recommend HPT.

Chapter 3

- Apply the criteria for timeliness and accuracy from the TSS.
- Report HPT to the FAIO.
- Template potential HVT and/or HPT.
- Recommend named/target areas of interest to the G-2 to support targeting.
- Participate in the dissemination and early planning process with the rules of engagement.

3-64. Coordinate with the collection manager section to ensure adequate intelligence collection to support targeting.

ASSISTANT DIVISION AIR DEFENSE OFFICER

3-65. The assistant division air defense officer's actions follow—
- Advise the commander and staff on the forward area air defense plan.
- Integrate Army airspace information from high to medium air defense and forward area air defense assets.
- Develop and maintains Army airspace utilization and situation.
- Request, maintain, and disseminate airspace control measures and restrictions.
- Synchronize friendly airspace usage with forward area air defense assets.
- Provide air and missile defense PIR to intelligence collection managers.
- Nominate active forward operating bases, forward air controller, and forward arming and refueling points.

AIR LIAISON OFFICER

3-66. The air liaison officer's action areas follow—
- Command the air support operation center and the corps and division TACP.
- Monitor the air tasking order.
- Advise the commander and his staff on the employment of air support assets.
- Review, coordinate, and assist with processing of preplanned air support requests/joint tasking air support requests.
- Coordinate redistribution of close air support resources.
- Coordinate approval of requests for immediate close air support.
- Provide air component input to analysis and plans.
- Receive, processes, exploits, and disseminates air intelligence.
- Provide intelligence support to EW operations.
- Provides air component PIR to intelligence collection managers.
- Coordinate supporting aircraft airspace requirements with airspace element.

AVIATION OFFICER

3-67. The aviation officers have the following responsibilities—
- Advise on employment of aviation.
- Coordinate aerial reconnaissance.
- Recommend airspace control measures for all aviation operations.

ELECTRONIC WARFARE OFFICER

3-68. The EW officer integrates the electronic attack in the targeting process and integrates electronic attack information requirements into the OPLAN, OPORD, and other planning products. The EW officer plans and coordinates electromagnetic activities. The EW officer interfaces between division and higher headquarters, the JFACC, multinational forces, and other components. The EW officer—
- Assists in coordinating electronic attack, electronic protection, and electronic support.
- Recommends to the commander's staff whether to engage a target with an electronic attack.

- Determines the electromagnetic requirements against specific HPT and HVT.
- Ensures electronic attacks meet the desired effect for targeting objective.
- Coordinates EW support and electronic attacks with the signals intelligence collection manager.
- Provides electronic attack requirements for airborne electronic attacks through the TACP.
- Serves as the jamming control authority for ground or airborne electronic attack.
- Coordinates with the chief of fires to integrate electromagnetic activities with overall fire support planning and into the targeting process.
- Coordinates with the chief of fires/fire support officer to prepare the fires annex for OPLAN and OPORD.
- Coordinates, prepares, and maintains the EW target list, electronic attack tasking, and requests.
- Identifies opportunities for effective targeting using electronic attacks.
- Assists the G-3 in coordinating EW requirements and tasking with the G-2, military intelligence unit commander, and other agencies as required.
- Coordinates with the assistant chief of staff, signal/signal staff officer to deconflict frequencies and the joint restricted frequency list with EW targets.
- Determines and requests theater army electronic attack support.
- Expedites electromagnetic activities reports to the targeting working group.

ENGINEER OFFICER

3-69. The engineer officer's actions are as follow—
- Advise on the obstacle and/or barrier plan.
- Advise on attack of targets with scatterable mines.
- Template potential HVT and/or HPT (mechanical breaching and minelayers).
- Help develop time-phase lines on the decision support templates and describe the effects of terrain on maneuver.
- War game and synchronize the effects of artillery scatterable mines.
- Develop HPTL, AGM, and BDA requirements with other members of targeting working group.
- Recommend HPT and named/target areas of interest to support the employment of artillery scatterable mines.
- Provide advice on environmental issues and coordinates with other members to determine the impact of operations on the environment.

PSYCHOLOGICAL OPERATIONS OFFICER

3-70. The psychological operations officer's responsibilities include—
- Advise the commander and unit staff on the military information support operations.
- Nominate military information support operations targets.
- Provide military information support operations input to the command targeting guidance.
- Coordinate military information support operations targeting with relevant sections such as fires, information operations, civil affairs, deception officer, and the information engagement officer.
- Conduct military information support operations planning.
- Evaluate the effectiveness of military information support operations with the intelligence directorate of a joint staff (J-2)/G-2 of the unit.

CIVIL AFFAIRS REPRESENTATIVE

3-71. The civil affairs operations staff officer—
- Advises on the effects of friendly operations on the civilian populace.
- Produces input to the restricted target list.
- Coordinates and provides situational awareness of the civil components to the IPB and targeting process.

Chapter 3

DIVISION, BRIGADE, AND BATTALION LIAISON OFFICERS

3-72. The actions of these liaison officers follow—
- Address the interests of the supporting commander as it relates to the supported commander's guidance on target nominations for shaping fires and operations.
- Coordinate with sending headquarters about targets added to the HPTL and the synchronization of the collection plan and AGM.
- Facilitate communication as soon as possible exchanging target information in sector, taskings, and coordination measures.

SIGNAL SUPPORT OFFICER

3-73. The signal support officer's actions follow—
- Manage information resources to support the commander's information requirements.
- Coordinate closely with the chief of staff, G-3 and other targeting working group members to synchronize information systems.
- Advise on the employment of information systems.
- Prepare the signal support annex to the OPORD and OPLAN.

OTHER PERSONNEL

3-74. During certain operations, personnel and agencies that will support the targeting process could include the following—
- Staff judge advocate.
- Deception officer and/or the information engagement officer.
- Air and naval gunfire liaison company.
- Army divisions and brigades coordinate Navy and Marine Corps support through an attached air and naval gunfire liaison company.

FIRES BRIGADE

3-75. Depending on the mission assigned to the corps or division, one or more fires brigades (FIB) may be included in the forces allocated for that mission. Fires brigades are normally assigned, attached, or placed in the operational control of a division. However, they may be attached or placed in operational control to a corps, JFLCC, a JTF or another Service component or functional component.

3-76. The FIB brings with it additional assets that augment the targeting capabilities of the corps or division. The FIB main CP fires cell is the primary organization responsible for integrating with the corps or division staff and for executing fires directed by the corps or division. The fires cell plans, coordinates, synchronizes, and integrates the fires, movement and maneuver, and protection warfighting functions for FIB operations.

3-77. The FIB fires cell includes operations and counterfire, target processing, fire control, lethal fires, nonlethal fires, air defense and airspace management, air support, and liaison elements. (See figure 3-2.) The operations and counterfire, target processing, and fire control elements form the nucleus of the FIB main CP current operations integrating cell. Each of the other elements in the fires cell assists these current operations core elements by providing additional expertise or dedicated manpower on an as needed basis. Conversely, scalable indirect fires, air defense and airspace management, air support, and liaison elements are the fires cell's primary contributors to the FIB main CP plans integrating cell, when that cell is activated. Elements and dedicated manpower in the fires cell assist with planning by providing expertise on an as needed basis.

Corps and Division Targeting

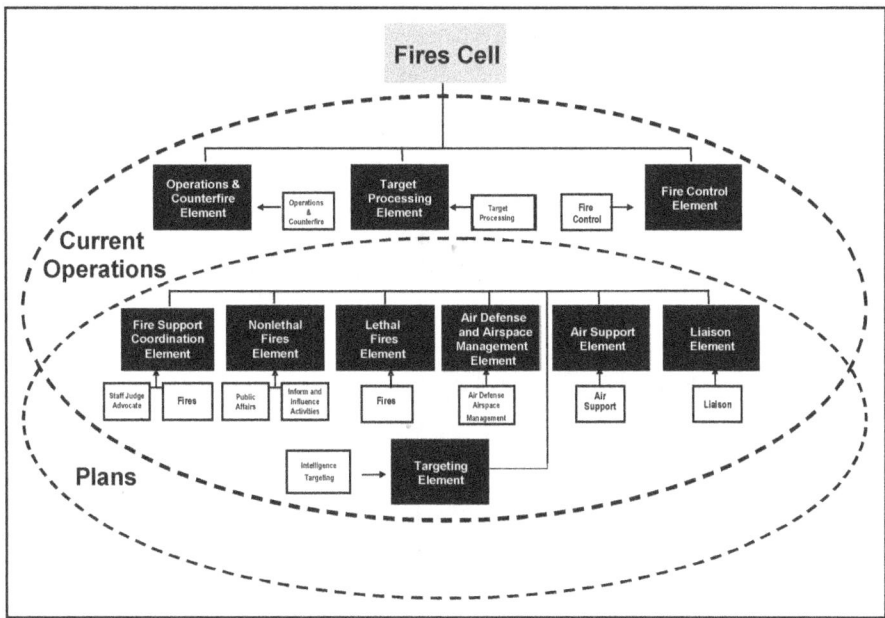

Figure 3-2. FIB fires cell and elements (example)

3-78. The following paragraphs provide an overview of the elements of the FIB fires cell.

Operations and Counterfire Element

3-79. Led by the FIB S-3, the FIB fires cell's operations and counterfire element's focuses not only on overall execution of both the FIB current operation but directs execution of its counterfire operation as well. The operations and counterfire element tracks and maintains situational understanding of all FIB assets. Among other responsibilities, the operations and counterfire element is responsible for—

- Execute fires in support of a division, corps, or other higher headquarters current operations (including special operations forces operating in the supported headquarters' area of operations).
- Act as the supported higher headquarters' force field artillery headquarters if so designated by the supported Commander.
- Position of assigned radars, meteorological sections, and supporting fire support-related units.

Target Processing Element

3-80. The FIB fires cell's target processing element is focused on reactive counterfire. The target processing element develops and nominates priority target sets, coordinates with the targeting (may be part of either the intelligence or fires cell) fire control, lethal and nonlethal fires elements, and participates in sessions of the targeting working group. Target processing element personnel ensure targets are prioritized and sequenced in current operations and future plans. Duties and responsibilities of the target processing element include—

- Recommend and update target acquisition coverage.
- Manage field artillery target acquisition assets and position areas.
- Orient field artillery target acquisition assets to ensure required coverage of the FIB air operations.

Chapter 3

Fire Control Element

3-81. The FIB fires cell's fire control element controls the delivery of tactical field artillery fires in support of current operations. It provides tactical fire control through automated weapon systems with manual backup and communications equipment.

Fire Support Coordination Element

3-82. The FIB fires cell's fire support coordination element supervises selected (normally the lethal and nonlethal fires elements and targeting elements) in planning, coordinating, synchronizing, and integrating the use of Army indirect fires, joint fires, and electronic attacks through the targeting process. This includes synchronizing physical attack and cyber/electromagnetic activities against the threat/adversary.

Lethal Fires Element

3-83. The FIB fires cell's lethal fires element (in concert with the nonlethal fires element) synchronizes the planning of fire support including Army indirect fires, joint fires, and electronic attacks to support the commander's intent through physical destruction, information and denial, enemy system collapse, and erosion of enemy will. The requirements for strike, counterfire, or fires in support of shaping operations will be given to the FIB in the form of mission orders. For example, if the division, corps, or other supported higher headquarters are conducting an attack to seize an objective or series of objectives, the FIB would likely receive tasks to isolate and reduce objectives, disrupt reinforcement, protect flanks, and interdict enemy artillery.

Nonlethal Fires Element

3-84. The FIB fires cell's nonlethal fires element (in concert with the lethal fires element) synchronizes the planning of fire support, including Army indirect fires, joint fires, and electronic attacks to support the commander's intent through physical destruction, information and denial, enemy system collapse, and erosion of enemy will. The fires cell synchronizes physical attack and cyber/electromagnetic activities against enemy/adversary capabilities. The FIB fires cell's nonlethal fires element is responsible for all aspects of scalable fires in support of FIB operations. This includes planning, coordinating, synchronizing, and integrating inform and influence activities (the FIB does not have an information engagement staff officer [S-7]) and the nonlethal aspects of EW activities for FIB operations.

Air Defense and Airspace Management Element

3-85. The FIB fires cell's air defense and airspace management element is designed to work with a division, corps, or theater airspace command and control element but is capable of limited independent operations should the FIB be employed independent of a division, corps, or other higher headquarters. The air defense and airspace management element is the principle FIB staff element that plans and coordinates airspace use by air and missile defense, Army aviation and unmanned aircraft system assets in support of FIB operations and then submits air control means requests to the division and corps airspace command and control elements for synchronization and deconfliction and further processing of air-space control means for inclusion in the airspace control order.

Air Support Element

3-86. The FIB fires cell's air support element consists of the Air Force TACP assigned or attached to the FIB.

Liaison Element

3-87. Duties and responsibilities of the liaison element include—
- Establish liaison with higher, adjacent, and supported units (as required).
- Exchange data and coordinating fire support across boundaries, when directed.

Targeting Element

3-88. The targeting element provides targeting support to the lethal fires element or works in concert with corps or division analysis and control element in target development based on mission. The targeting element also prepares recommendations for FIB targeting working group sessions and implements the resulting decisions through targeting guidance.

This page intentionally left blank.

Chapter 4
Brigade Combat Team and Battalion Task Force Targeting

The brigade combat team (BCT) commander synchronizes warfighting functions within their area of operations. Fires, intelligence, movement, and maneuver are warfighting functions integrated into the planning and operations to accomplish the BCT missions. The focus of the targeting effort comes from—

- The division plan and/or order.
- The BCT mission statement.
- The BCT commander's intent.

The BCT battle is essentially the division's close combat. The targeting decisions at a higher headquarters affect targeting decisions at subordinate headquarters. The BCT staff uses the targeting products of the division. Division level tasking is integrated into the BCT targeting process. BCT targeting addresses assets under BCT control. The high-payoff target list (HPTL) and attack guidance matrix (AGM) at BCT and task force are normally more detailed and focused. They provide the information the sensor or observer and a weapon system require to identify and attack high-payoff targets (HPT).

Targeting at the BCT and battalion task force level is frequently not as formal as at higher headquarters. The task force may not develop its own formal HPTL or AGM in the format presented in this manual. However, the concept of the targeting process is still valid and useful at task force level. The task force uses or modifies the existing BCT HPTL, AGM, and other targeting products. At the task force level, the HPTL is developed through war gaming. HPT matrix replicates the task force decision support template and addressed in a synchronization matrix. The synchronization matrix identifies the responsible observers with each HPT, designation of weapon systems, and activation of associated control measures such as fire support coordination measures (FSCM)/airspace coordinating measures. The synchronization matrix addressing friendly and enemy actions may be posted on operational graphics supporting the task force operation order (OPORD) or operation plan (OPLAN). A more formal representation of this information may be developed in a separate HPTL, collection plan, and fire support execution matrix. The focus of the decide function of the targeting process at task force level is to provide observers with critical information. The observer detects targets and passes target acquisition reports to the fire direction center of the weapon systems so that the maneuver receives timely and effective fire support.

FUNCTIONS

4-1. Targeting functions at BCT and task force level include the following—
- Develop the HPTL.
- Develop attack guidance.

Chapter 4

- Establish target selection standards (TSS).
- Nominate targets to higher headquarters.
- Synchronize the intelligence, surveillance, and reconnaissance (ISR) plan.
- Synchronize maneuver and fire support.
- Integrate countermobility, mobility, and survivability operations.
- Receive and evaluate battle damage assessment (BDA).
- Monitor fire support systems and ammunition.
- Develop and synchronize the ISR plan with the fire support plan. (Focus on positioning observers early to support the top-down fire plan.)

PLANNING CONSIDERATIONS

4-2. The fast-paced, ever-changing nature of the battlefield at BCT and task force levels presents challenges to the targeting process, including—
- BCT battle rhythm affects when and where targets will be acquired.
- Targets are generally highly mobile.
- The BCT has limited assets with which to detect and attack long range targets, especially moving targets.
- Planning time is limited, and planners are executers.
- Planning is primarily focused on current operations out to the next 36 hours.

4-3. Planning considerations at the BCT and task force levels are very similar. Plans must be simple, but contain sufficient detail so that subordinate units can execute with precision and vigor. Rehearsals are critical for success on the battlefield. Planning time must be allocated for rehearsals. Rehearsals clarify the fire plan for observers, sensor operators, weapon system managers, and the maneuver units they support. Rehearsals facilitate the synchronization of maneuver with fires and other warfighting functions. Rehearsals result in an improved understanding of the situation, commander's intent, concept of operations, and tasks to subordinate units throughout the brigade sector of operations.

4-4. There is normally not enough planning time available for the brigade fire support officer (FSO)/battalion FSO to wait for subordinate elements to forward targets for inclusion in the fire support plan. Top-down fire planning overcomes this lack of planning time. Fire support plans are disseminated to subordinate levels as early as possible and contain the following—
- Commander's intent and concept of fires.
- Targeting guidance.
- HPTL and AGM.
- Specific tasking for BCT targets.
- FSCMs, air coordination measures, and other graphic control measures for integration and synchronization with task force plans.

4-5. A fire support execution matrix is often used to disseminate this information.

4-6. The brigade/battalion FSO establishes a reasonable cutoff time for submitting routine changes to the target list before the start of combat operations. Targeting is a continuous process, and emergency and critically important changes will be accommodated anytime. However, the time for routine changes must be limited to allow time to finalize, disseminate, and rehearse the fire support plan.

4-7. Targeting functions at task force level rely heavily on the targeting products from BCT. The targeting working group must understand the BCT commander's targeting guidance, to include the following—
- Criteria for attack and engagement.
- HPT.
- Any constraints during each phase of the battle.

4-8. The targeting working group must know—
- What targets are planned in the task force area of operations?

Brigade Combat Team and Battalion Task Force Targeting

- What responsibilities the team has for BCT targets?
- What targeting detection and delivery assets are allocated to the task force?

4-9. For example, assets could include all of the following—
- Close air support sorties.
- Tactical unmanned aircraft systems.
- Combat observation and lasing teams (COLT).
- Army aviation support.
- Priorities of fires.
- Allocation of special types of ammunition, such as family of scatterable mines and dual-purpose improved conventional munitions.

4-10. The targeting working group must know the FSCM in effect during each phase of the battle. The cutoff time for submission of changes to the target list and the BCT rehearsal time are also needed.

TARGETING WORKING GROUP BATTLE RHYTHM

4-11. The targeting working group is a critical event in the BCT battle rhythm. The timing serves to nest the BCT battle rhythm into the higher headquarters targeting process. Targeting working group sessions must be effectively integrated into the BCT battle rhythm and nested within the higher headquarters targeting cycle to ensure that the results of the targeting process focuses, rather than disrupt operations. (See figure 4-1.) Thus, task organization changes, modifications to the ISR plan, target nominations that exceed organic capabilities, air support requests, and changes to the HPTL and specified fire support, inform and influence activities, and electronic warfare (EW) tasks all must be made with full awareness of time available to prepare and execute.

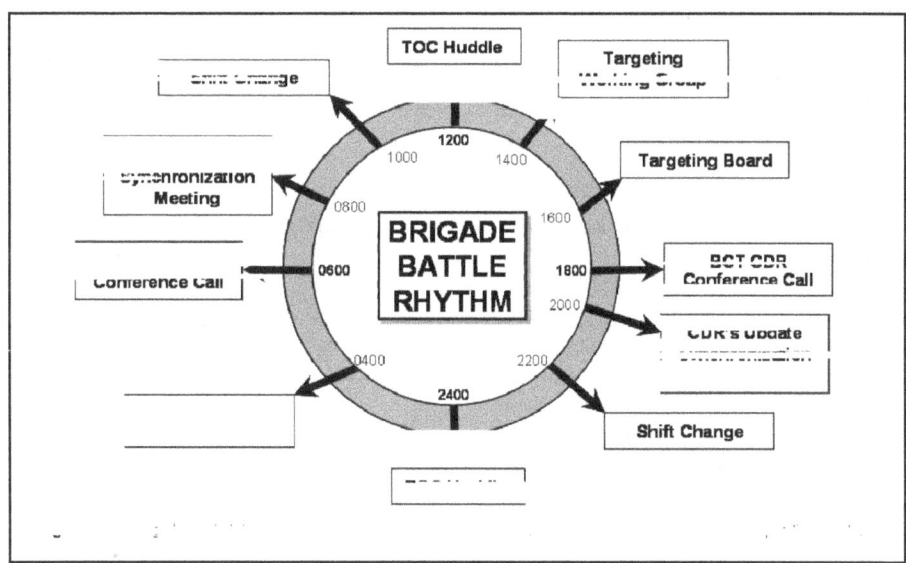

Figure 4-1. Brigade battle rhythm (example)

4-12. The timing of targeting working group sessions is critical. While the time-focus for BCT level sessions of the targeting working group is normally 24 to 36 hours out, the BCT employment of "reachback" assets and certain targeting decisions, such as target nominations and air support requests, must be planned in advance and in conjunction with the division, corps, theater army, and the joint air tasking cycle. For these reasons, the BCT targeting focus is 24, 48, and 72 hours. However, commanders must choose a targeting cycle based on the pace of operations in their area of operations. In stability operations the targeting timeline may be extended to focus as much as one month or more in the future. The brigade FSO also schedules internal fires cell targeting meetings so fire support, inform/influence activities, and EW activities related target nominations arrive within the BCT and higher echelon target nomination windows.

4-13. Experience in the BCT has shown the benefits of two targeting working group sessions daily at the main command post (CP). A preliminary session facilitated by the fires cell ensures the effects of scalable fires and meet the BCT commander's guidance and intent. The brigade FSO, fires cell planners, and targeting working group assess ongoing targeting efforts, and ensure air support requests with target nominations are processed through higher headquarters to meet division, corps, theater army, and joint task force targeting timelines. The second session is generally more formal than the first and is focused on updating the commander, gaining new guidance, and obtaining approval of planned and proposed targeting actions. Targeting working group sessions should be the minimum length required to present targeting information, situation updates, provide recommendations, and obtain decisions.

BCT TARGETING WORKING GROUP MEMBERSHIP

4-14. The targeting working group is a grouping of predetermined staff representatives concerned with targeting who meet to provide analysis, coordinate, and synchronize the targeting process, and provide recommendations to the targeting board. The targeting working group focuses and synchronizes the BCT combat power and resources toward finding, tracking, attacking, and assessing HPT. The following personnel should normally attend sessions of the targeting working group—

- Brigade FSO (leads the working group).
- BCT operations staff officer (S-3) (alternate lead).
- BCT intelligence staff officer (S-2) representative.
- BCT collection manager.
- Information engagement staff officer (S-7).
- EW officer.
- Psychological operations noncommissioned officer.
- BCT civil affairs staff operations officer (S-9) representative.
- Fires cell targeting officers.
- Brigade legal section.
- Organic fires battalion S-2/S-3 (if available).
- Combat observation and lasing and fire support team chiefs (if available).
- Fires cell representatives from the maneuver battalions, the reconnaissance squadron, and the brigade special troops battalion (if available).
- Military intelligence company commander/collection manager.
- Air Force air liaison officer/tactical air control party (TACP) representative.
- Air defense airspace management/brigade aviation element representative.
- Special operations forces representative.
- Reinforcing unit liaison officers.
- Naval gunfire liaison officer/naval surface fire support representative.
- BCT engineer officer (assistant brigade engineer/engineer liaison officer in his absence).
- BCT chemical, biological, radiological, and nuclear officer.
- Weather officer.
- BCT logistics staff representative.
- BCT signal staff representative.
- Joint, interagency, multinational representatives as needed.

BCT TARGETING BOARD MEMBERSHIP

4-15. The targeting board is a temporary grouping of selected staff representatives with delegated decision authority to provide targeting decision recommendations for command approval. When the

process or activity being synchronized requires command approval, a board is the appropriate forum. The targeting board usually includes—
- BCT executive officer (chairs the board).
- BCT S-3 (alternate chair).
- BCT intelligence staff officer.
- BCT signal staff officer.
- BCT information engagement staff officer.
- BCT civil affairs operations staff officer.
- Fire support coordinator.
- Brigade FSO.
- BCT engineer.
- Air defense airspace management/brigade aviation element officer.
- EW officer.
- Air liaison officer.
- Fires cell targeting officers.
- Psychological operations detachment commander.
- Civil affairs unit leader.
- Brigade legal section.
- BCT logistics staff (sustainment cell) representative.
- Organic fires battalion S-2/S-3 (if available).
- Combat observation and lasing and fire support team chiefs (if available).
- Representatives from the maneuver battalions, the reconnaissance squadron, and the brigade special troops battalion (if available).
- Military intelligence company commander/collection manager.
- Special operations forces representative.
- Reinforcing unit liaison officers.
- Naval gunfire liaison officer/naval surface fire support representative.
- Assistant brigade engineer/engineer liaison officer.
- Joint, interagency, multinational representatives as needed.

BCT TARGETING RESPONSIBILITIES

4-16. The following provides a brief overview of targeting specific responsibilities for selected BCT personnel.

BCT Commander

4-17. The BCT commander directs the targeting effort. Before preparation of formal running estimates, guidance and direction is provided on the following—
- State the expectation for the command.
- Establish the what, when, and why to accomplish the mission.
- Intent for shaping the battlefield in terms of time and space.
- Identify the critical enemy vulnerability that will lead most directly to accomplishing the mission.
- Stress the time and place as critical element during the battle.
- State the desired end state respecting time, force, enemy, and terrain.
- Establish guidance for units having priority of fires.
- Identify high-value targets and state the desired effects.
- BDA requirements.
- State the rules of engagement and commander's intent.

Chapter 4

4-18. With this information, the staff prepares formal estimates. After presentation of these estimates, the commander refines the previous guidance. Who approves the HPTL developed by his brigade FSO, S-3, and S-2, or amends the list and approves it. HPT that cannot be acquired or attacked with BCT assets are forwarded to the division targeting working group for consideration. The brigade FSO develops the attack guidance and submits it to the commander for approval. As the battle progresses and more information become available, the commander may have to change his guidance to react to changes.

BCT Deputy Commander

4-19. The BCT deputy commander is second in command of the BCT and is prepared to assume command at any time necessary. The role, responsibilities, and authority vary based on the commander's desires, the BCT mission, and the scope and complexity of the operations conducted by the brigade. The BCT commander delegates responsibility to his deputy commander for specific areas and/or functions, and the authority necessary to control them in order to extend the commander's span of control.

BCT Executive Officer

4-20. The BCT executive officer is the commander's principal staff leader. Who directs, coordinates, supervises, trains, and synchronizes the work of the staff, ensuring efficient and prompt staff actions. The commander normally delegates executive management authority (equivalent to command of the staff) to the executive officer for the coordinating and special staff. These decisions often include modifications to targeting products. Decisions normally include approving or modifying—

- HPTL/refinements.
- Targeting synchronization matrix.
- BCT focus for fire support.
- Air support request/joint air support request for air interdiction and close air support or the refinement of those previously submitted.
- Attack plan/refinements.
- Fragmentary orders.
- Measures of performance (MOP) and measures of effectiveness (MOE) for scalable fires.
- Fire support tasks.
- Inform and influence tasks.
- EW tasks.

BCT Intelligence Officer

4-21. The S-2 is responsible for preparing the ISR plan and maintains information on the current enemy situation. The role provides assessment of possible enemy actions, provides analyses, and identifies targets based on the BCT commander's guidance. Specific targeting responsibilities include—

- Developing target arrays.
- Providing enemy capabilities and projected courses of action.
- Providing intelligence preparation of the battlefield (IPB) products to the targeting working group.
- Developing high-value targets (HVT) in coordination with staff.
- Determining which HPT can be acquired with organic assets.
- Developing support requests for acquiring HPT beyond the capabilities of organic assets.
- Coordinating the collection and dissemination of targeting information with the targeting officers in the fires cell.
- Developing and supervising implementation of the ISR plan.
- Advising the S-3 about assessment collection capabilities.
- Coordinating with the brigade FSO for indirect fires to support the ISR plan.
- Participate in the dissemination and early planning process of rules of engagement.

4-22. The BCT S-2 must inform other BCT staff personnel, the field artillery S-2, and task force S-2 of the following—
- Target arrays.
- Enemy capabilities and projected course of action (COA).
- The civilian situation (in the absence of the S-9).
- HVT.

4-23. To do this, the S-2 leads the staff IPB, by using the tools of target value analysis, and applies what he knows about the enemy situation. This provides a base of HVT that is adjusted according to current enemy dispositions and composition.

4-24. After the HPTL is approved, the S-2 determines which targets can be acquired with organic, attached or assets in support of the BCT. Targets that cannot be acquired at brigade level are translated into requests for information from higher headquarters.

4-25. The BCT S-2 coordinates with the fires battalion S-2, targeting officer, and task force intelligence officers for the collection and distribution of targeting information. This includes production of the TSS matrix for target acquisition assets supporting the BCT. The S-2 also plans and supervises an aggressive collection effort focused on the BCT HPT, priority intelligence requirements (PIR), and intelligence requirement.

4-26. The S-2 must inform the other members of the staff when major changes in the tactical situation warrant reevaluation of the HPTL. After the S-3 and brigade FSO state requirements for timeliness and accuracy, the S-2 must translate these into collection tasking. The S-2 must work closely with the brigade FSO and S-3 to decide which targets are best suited for coordinated attack. A coordinated attack may involve a combination of a wide array of attack methods, to include—
- Destructive methods.
- Electronic monitoring.
- Offensive EW.
- Deception to enhance the effectiveness of the attack.

4-27. The S-2 must also advise the S-3 on BDA collection capabilities.

4-28. The S-2 is relied on significantly by the S-3 to assist him in the development of a comprehensive ISR plan. The plan must tell commanders what they need to know in time for them to act. It is commander oriented and commander directed. The ISR plan will answer the commander's PIR by tasking appropriate assets to gather information. It will also ensure that observers are focused on designated named area of interest to facilitate the targeting effort. The BCT S-2 and the brigade FSO will coordinate indirect fires planned to support ISR assets. (See Field Manual Interim (FMI) 2-01 for additional information.)

BCT Operations Officer

4-29. The BCT S-3 must work closely with the BCT S-2 and brigade FSO to prioritize the HPTL before its approval by the commander. Priorities should address the following—
- When the targets should be engaged.
- The desired effect on the target.
- Those target types that should be attacked immediately.

4-30. The decision to designate a target type for immediate attack is especially critical. Assets are diverted from a mission in progress to attack that target type.

4-31. The S-3 is responsible for giving a detailed interpretation of the commander's concept of the operation to all personnel engaged in brigade-level targeting. The guidance that results from this interpretation must specify the targets that the commander feels are most important and the targets that pose the greatest threat to the mission. The S-3 should specify the desired effects on the target when they are different from those recommended by the brigade FSO or S-2. The where or when HPT should be attacked for the greatest benefit to the friendly operation. The S-3 or brigade FSO must coordinate with division for

Chapter 4

those targets that are important to the BCT. Although the emphasis is on HPT, other targets of lower priority also may be attacked.

4-32. The BCT S-3's specific targeting responsibilities include—

- Working with the S-2 and brigade FSO to prioritize the HPTL before approval by the commander.
- Synchronizing fires, intelligence, and airspace requirements with maneuver operations.
- Determining the targets to be attacked immediately and desired effects.
- Providing a detailed interpretation of the commander's concept of the operation.
- Providing guidance about which targets are most important to the commander.
- Deciding when and where targets should be attacked.
- Periodically reassessing the HPTL, TSS, AGM, targeting synchronization matrix, MOP, MOE, and fire support tasks with the brigade FSO and S-2.
- Determining with the brigade FSO and S-2 if an attack resulted in the desired effects or if additional attacks are required.
- Coordinate the dissemination and early planning process of rules of engagement.

Fire Support Coordinator

4-33. The BCT organic fires battalion commander is the fire support coordinator. The fire support coordinator is the BCT commander's primary advisor for planning and employing the field artillery assets. Additional responsibilities include integrating all scalable fires for his mission.

Brigade Fire Support Officer

4-34. The brigade FSO is the senior field artillery staff officer at brigade level. This position is responsible for all scalable fires planning and execution.

4-35. The brigade FSO plans and coordinates the fires warfighting function for BCT operations. The position works closely with the executive officer and S-3 to ensure mutual understanding of all aspects of fire support assessment, planning, preparation, and execution for BCT operations. Responsibilities are to assist as needed in planning, during the transition to execution, and flowing smoothly into execution. The brigade FSO finalizes the attack guidance formulated by the BCT commander and chairs the targeting working group. His targeting specific actions are—

- Overseeing overall targeting execution.
- Ensuring all aspects of targeting are addressed and understood during the targeting process (task, purpose, location sensor/back-up, fire mission thread, rehearsal, delivery asset, and assessment).
- Developing and updating targeting products including fire support tasks, HPTL, TSS, AGM, targeting synchronization matrix, sensor-shooter matrix, MOP, and MOE for scalable fires.
- Conducting assessment in conjunction with the S-2 and S-3.
- Preparing the fire support execution matrix.
- Ensuring subordinate battalion FSO and fires cells fully understand target execution responsibilities and planning fire support in accordance with the BCT commander's top down fire plan.
- Establishing target refinement standards to facilitate completion of the fire support plan prior to execution.
- Consolidating target refinements and planned targets from the subordinate battalion/squadron FSO, fires cells, and resolving duplications.
- Providing target refinement to higher headquarters for established division, corps, and theater army targets.
- Coordinating requests for additional fire support from subordinate battalion/squadron FSO and fires cells.
- Coordinating support for subordinate unit attack requirements.

- Coordinating suppression of enemy air defense (SEAD), joint suppression of enemy air defense (J-SEAD), and joint air attack team operations.
- Receiving assessment reports and, with the S-2 and S-3, determining if an attack resulted in the desired effects, or if additional attacks are required.
- Formulating the re-attack recommendation.
- Ensuring target nominations are validated, processed, and updated to achieve the desired effect using joint air assets.
- Coordinating with the air component air liaison officer on use of tactical air assets.

Targeting Officers

4-36. The targeting officer in the fires cell facilitates the exchange of information among the military intelligence company's analysis and control team, BCT S-2, and subordinate fires cells. These responsibilities are similar to the field artillery intelligence officer at the division and corps. The targeting duties include—
- Helping the BCT S-2 to develop the ISR and target acquisition plans.
- Helping to provide staff supervision of target acquisition assets organic to, attached to, or under control of the BCT.
- Coordinating with the BCT S-2 for target acquisition coverage and processing of HPT.
- Producing the targeting synchronization matrix for target acquisition assets supporting the BCT.
- Conducting target coordinate mensuration when applicable.
- Conducting munitions effects analysis (weaponeering) when applicable.
- Conducting collateral damage estimation when applicable.
- Developing, recommending, and disseminating the AGM, MOP, and MOE to subordinate elements.
- Developing, recommending, and disseminating approved fire support tasks to subordinate elements.
- Managing target lists for planned fires.
- Coordinating and distributing the restricted target list in coordination with the brigade FSO.

Air Force Air Liason Officer

4-37. The air component air liaison officer's targeting actions are—
- Monitoring execution of the joint air tasking cycle.
- Advising the commander and staff about employment of air assets.
- Receiving, coordinating, planning, prioritizing, and synchronizing immediate requests for close air support.
- Providing air component input to analysis and plans.

Information Engagement Staff Officer

4-38. The S-7 is responsible for the overall planning, preparation, execution, and assessment of information tasks for the BCT. The position targeting responsibilities include—
- Synchronizing appropriate aspects of inform and influence activities with the fires, maneuver, and other warfighting functions.
- Assessing enemy vulnerabilities, friendly capabilities, and friendly missions.
- Nominating inform/influence activities targets for attack.
- Briefing deception operations.
- Providing operation security measures.
- Synchronizing Army information tasks.

Electronic Warfare Officer

4-39. The EW officer's targeting responsibilities include—
- Determining HPT to engage with electronic attack.
- Submitting air support requests and airspace control means requests for EW aircraft support.
- Recommending EW methods of target engagement.
- Planning and coordinating tasking and requests to satisfy electronic attacks and EW support requirements.
- Assisting the S-2 with the electronic portion of IPB.
- Identifying threat electronic attack capabilities and targets.

Psychological Operations Noncommissioned Officer

4-40. The psychological operations noncommissioned officer targeting responsibilities include—
- Specifying military information support operations targets during the target nomination process and recommending them to the targeting team.
- Providing assessments of military information support operations actions and programs.
- Identifying and nominate targets for the restricted target list and no-strike list.
- Coordinating and deconflicting military information support operations targeting with planners.
- Synchronizing and deconflict military information support operations with subordinate elements.
- Providing military information support operations relevant information.

Civil Affairs Officer

4-41. The S-9's targeting responsibilities include—
- Providing advice on the affects of friendly actions on the civilian populace.
- Producing input to the restricted target list.
- Providing assessments of the effectiveness of civil affairs operations.

Brigade Judge Advocate

4-42. The brigade judge advocate's targeting responsibilities include—
- Analyzing the operations relative to the rules of engagement, United States laws, existing host nation law, and international law.
- Analyzing the nominated or potential target under the law of war.
- Analyzing the plans for detention operations can include evaluation for potential future criminal prosecution of a target, site exploitation, and evidence preservation.
- Identifying the need for potential legal support to operations.
- Provide interpretations of the rules of engagement.

Liaison Officers

4-43. Liaison officer targeting responsibilities include—
- Addressing concerns of their respective commanders. Submitting and explaining the significance of target nominations to support their unit's operations.
- Providing feedback to their respective commanders on which targets are added to the HPTL and how they are synchronized with the ISR plan and AGM.
- Providing feedback to their commanders on target nominations made to higher headquarters.
- Informing their organization of higher level targets that fall into their sectors along with the tasking and coordination measures involved.
- Providing supported unit with required targeting information from the supporting or subordinate unit.

BCT FIRES CELL

4-44. The BCT fires cell is led by the brigade FSO. It is staffed by members that have expertise that is integral to the fires warfighting function. It has resources to plan for future operations from the main CP and to support current operations from the tactical CP (when deployed). Additionally the section has the limited capability to provide coverage to the command group and the deputy command group when deployed. Fires cell staff are assigned to the following elements within the fires cell of the main CP and current operations cell of the tactical CP—

- Lethal fires element (main CP).
- Nonlethal fires element (main CP).
- The TACP operates from the main CP with selected personnel deployed with the tactical CP when deployed.
- Tactical CP fires element. Selected personnel from the lethal and nonlethal fires elements deploy with the tactical CP when the tactical CP is deployed, otherwise they are part of main CP fires cell.

4-45. The fires cell of the main CP expands their functions to include the follow— inform and influence activities, civil affairs operations personnel, an enhanced air component TACP, and collocation of the brigade legal section. The result is a multi-functional organization with much improved capability including—

- The improved capability of planning, integrating, and synchronizing activities for BCT operations.
- The improved ability to integrate available capabilities into targeting.
- The capability of managing counterfire operations when required.
- Improved joint fires connectivity.

4-46. All elements work from the main CP if the tactical CP is not deployed. Selected personnel from the main CP fires elements make up the fires element of the tactical CP (the entire tactical CP is the BCT current operations cell) when the tactical CP is deployed. The rest remain at the main CP. The BCT TACP collocates with the fires cell in the main CP (but is sufficiently robust that a selected portion of it can be deployed with the tactical CP). As mission, enemy, terrain and weather, troops and support available, time available, civil considerations (METT-TC) dictates, the fires cell in the main CP can be augmented by other Service components or joint resources and assets, including those for inform and influence activities, and related activities as needed.

4-47. The S-7, S-9, and public affairs officer plan, coordinate, integrate, and synchronize all aspects of inform and influence activities to support BCT operations. The S-7, S-9, and public affairs officer conduct all aspects of their duties to support BCT operations, to include supporting the brigade FSO with the integration of their capabilities into the targeting process. They are part of the targeting working group that integrates inform and influence activities into the targeting process. The fire support organizations at battalion/squadron level and below coordinate fires, to include organic mortars and any additional fire support allocated by the BCT fires cell, for delivery on time and on target. Together the BCT fires cell and the fires cells of subordinate BCT organizations enable the BCT to conduct operations to protect the force and shape the battlefield.

4-48. The fires cell is the centerpiece of the BCT targeting architecture, focused on both lethal and nonlethal target sets. The fires cell thus collaboratively plans, coordinates, and synchronizes fire support, aspects of inform and influence activities (such as artillery and air delivered leaflets) in an integrated fashion with the other warfighting functions to support BCT operations. The targeting working group brings together representatives of all staff sections concerned with targeting. It synchronizes the contributions of the entire staff to the work of the fires cell. The brigade legal section is colocated with the fires cell in order to provide legal review of plans, targeting and orders. The fires cell coordinates and integrates joint fire support into the BCT commander's concept of operations. Primary functions of the fires cell includes the following—

- Planning, coordinating, and synchronizing fire support for BCT operations.
- Working with the S-7, S-9, public affairs officer, and brigade judge advocate to integrate fires, including appropriate aspects of inform and influence activities into the BCT targeting process.

- Collaborating in the IPB process.
- Coordinating the tasking of sensors during development of the ISR plan with the BCT S-2, the military intelligence company commander (as needed), and the reconnaissance squadron to acquire targets.
- Participating in the BCT military decisionmaking process (MDMP).
- Briefing the BCT commander on the fire support plan.
- Disseminating the approved plan to BCT fire support organizations, the fires battalion, the division's fires brigade and the division and corps fires cells.
- Participating in the BCT targeting process.
- Ensuring battalion fires cells plan fires in accordance with the BCT commander's guidance for current and future operations.
- Preparing the fires portion of the BCT OPORD that describes the concept/scheme of fires to support BCT operations.
- Managing the establishment of and changes to FSCM.
- Coordinating maneuver space for the positioning of field artillery assets.
- Coordinating airspace requirement with the air defense and airspace management/brigade aviation element.
- Submitting airspace control means requests to integrate airspace requirement.
- Coordinating clearance for attacks against targets (clearance of fires).
- Coordinating assessment.
- Coordinating requests for additional fire support to include joint fires.
- Providing input to the common operational picture to enhance situational understanding.

Lethal Fires Element (Main CP)

4-49. A *fires element* is a component of the fires cell. A fires element is normally collocated with its parent fires cell but may operate within another CP cell. The principal function of the lethal fires element is planning scalable fires for future BCT operations and targeting. The BCT main CP facilitates collaboration of fires with the other warfighting functions. The lethal fires element prepares scalable fires inputs and products used in the MDMP and targeting process. On adoption of a COA, the element produces and disseminates the fire support portions of the BCT OPLAN/OPORD. The lethal fires element prepares recommendations for BCT targeting working group sessions and implements the resulting decisions through the targeting guidance in fires computer systems. Leveraging the ISR assets available at the main CP, the element plans and executes the scalable fires portion of BCT shaping operations. Functions of the fires cell lethal fires element include the following—

- Providing lethal fires input to the ISR plan.
- Developing the fire support concept for each COA.
- Developing no-strike lists and identifying FSCM.
- Developing/refining targeting guidance for each COA.
- Developing target criteria for input into computer systems for each COA.
- Producing the HPTL, TSS, and AGM, targeting synchronization matrix and lethal fires tasks for the BCT OPLAN/OPORD.
- Preparing products for the targeting working group.
- Developing fire support related MOP and MOE for BCT assessment.
- Implementing targeting guidance in computer systems.
- Providing reactive counterfire guidance and radar deployment instructions to the organic fires battalion.
- Updating/purging targeting files.
- Assisting the operations element in clearance of fires when required.
- Planning and executing lethal fires for BCT shaping operations as directed.

Lethal Fires Element Personnel and Their Responsibilities

Assistant Brigade Fire Support Officer

4-50. The assistant brigade FSO is a field artillery officer who assists the brigade FSO in all duties and is prepared to assume the brigade FSO responsibilities. The assistant brigade FSO and the brigade FSO split responsibilities between the main CP and the tactical CP when both CPs are deployed. The BCT commander and brigade FSO determine who operates from which location and the manning and equipment needed to support the operation. The command group or the assistant brigade FSO serves as a shift leader in the fires cell element when not deployed with the tactical CP.

4-51. The assistant brigade FSO participates in the war gaming process to develop the HPTL and AGM. The BCT fire support execution matrix is completed by assistant brigade FSO. The assistant brigade FSO coordinates with the task force FSO to ensure task force plans—

- Meet the BCT commander's guidance.
- Avoid unplanned duplication.
- Use all assets assigned to the task force.
- Develop observer overlay with assigned observers for all BCT targets as required.

4-52. The assistant brigade FSO coordinates the attack of targets by all fire support weapons system. A critical task consists of positioning and controlling the BCT observation assets with the BCT S-2. The position advises the commander, executive officer, and S-3 on the following—

- The ability of the fire support systems to defeat HPT and other designated targets.
- The best means of attack.
- The best type of munitions to achieve the commander's desired results.

4-53. Once the HPTL is approved, the assistant brigade FSO ensures that fire planning and fire support requests are processed according to the BCT commander's guidance. The assistant brigade FSO informs the fires battalion and task force FSO of the target types designated HPT and targets that must be processed quickly. The position is responsible for developing the AGM.

Targeting Officers

4-54. The BCT fires cell is organized with two targeting officers. These officers serve as advisors to the brigade commander, brigade FSO, and the brigade staff on all fire support and ISR assets available to the brigade. This assists the brigade staff in maximizing fire support coverage and synchronizing effects on the battlefield in order to achieve the commander's intent and end state.

4-55. Targeting officers work with the plans section, brigade FSO, and the targeting working group during the MDMP to determine which targets need to be engaged and the desired effects for each engagement in order to achieve the commander's intent. These individuals produce the targeting and assessment guidance to be distributed with the BCT OPLAN/OPORD. The targeting team develops the follow tools—

- HPTL.
- AGM.
- TSS.
- Targeting synchronization matrix.
- Fire support related MOP and MOE for BCT assessment.
- Coordinate with the TACP for air support request and target area deconfliction.

4-56. The targeting officers collect, analyze, process, produce, and disseminate targeting information and products. The information and products are necessary for the employment of coordinate seeking precision guided munitions. These individuals' role is vital to synchronizing all target acquisition assets attached, organic, or under control of the BCT. Providing reactive counterfire guidance, radar deployment instructions are essential to the organic fires battalion S-2. The BCT acquisition systems and assets available through reach-back are critical to locating HPT for attack. These individuals assist in collateral damage estimation recommendations for the BCT commander and staff.

4-57. The targeting officers are integral to fusing the BCT fires cell with the intelligence apparatus. The ability to gain and maintain information superiority between the fires cell, the BCT S-2, and the FA battalion S-2 allows commanders' to execute decisions and maintain the initiative. The role of the targeting officer is similar to the functioning of the field artillery intelligence officer (FAIO). In this capacity, these individuals help the S-2 and the brigade FSO determines specific target vulnerabilities. The enemy's current state of vulnerabilities achieved by defeat is information that must become knowledge among commanders and staffers. The ability to identify these functions or capabilities that were defeated enhances the operational picture. The targeting officer, in coordination with the brigade FSO, consolidates and distributes the target list, restricted target list, and no-strike list. During operations, these individuals monitor compliances with the restrictions and report incidents where the restrictions may have been violated.

4-58. The targeting officers provide recommendations to the targeting working group on updating targeting priorities. These individuals prepare products for the targeting working group as directed by the brigade FSO. They direct updating and purging of targeting files as required. The targeting officers ensure that interoperability is maintained with collection assets of the BCT. The reactive counterfire guidance and radar deployment instructions are critical to the brigade FSO, plans section, targeting working group, and all subordinate units.

4-59. The brigade headquarters is organized with a main and a tactical CP. These targeting officers provide the targeting experience to conduct 24 hour brigade operations and targeting expertise in both CP. Also, this capability allows the commander the option of assigning the responsibilities of lethal and nonlethal targeting and/or assessment amongst the targeting officers.

Fires Cell Operations Noncommissioned Officer

4-60. The fires cell operations noncommissioned officer is the senior enlisted assistant to the brigade FSO. This individual must understand and actively participate in the MDMP and production of the OPLAN/OPORD. The position serves as shift leader in the fires cell; either at the main CP or tactical CP. Major responsibilities include the following—
- Ensuring that the fires cell is fully manned for 24-hour operations and all of its equipment is fully functional.
- Performing fires cell digital network management and troubleshooting to ensure internal and external connectivity.
- Supervising the enlisted personnel in the fires cell and processing administrative matters pertaining to the fires cell.
- Managing fires cell situational understanding input to the BCT common operational picture.
- Managing FSCM and ensuring they are accurately tracked throughout the BCT.
- Coordinating with the air defense and airspace management/brigade aviation element for airspace requirements necessary to integrate fires with other airspace user.
- Preparing required reports in accordance with BCT standing operating procedures.
- Maintaining files and documents.
- Developing and enforcing the fires cell standing operating procedures.

Fire Support Noncommissioned Officers

4-61. The fire support noncommissioned officers function as enlisted assistants to the brigade FSO and the assistant brigade FSO. One or the other may deploy with the brigade FSO and command group 1. The fire support noncommissioned officer assists the shift leaders as needed in either the fires cell operations element (tactical CP) or fires cell plans and targeting element (main CP) to enable 24-hour operations when not deployed with the brigade FSO.

Target Analyst/Targeting Noncommissioned Officers

4-62. The target analyst and the targeting noncommissioned officers, together with the targeting officers, provide a 24-hour capability to plan and coordinate targeting operations. Their primary responsibilities include the following—
- Operating and maintaining the targeting computer systems.
- Maintaining the targeting common operational picture display.

- Maintaining the target production display.
- Updating and purging targeting files as directed by the targeting officers.
- Ensuring targets that are acquired are processed to the appropriate fire support assets in accordance with the targeting synchronization matrix.
- Ensuring essential voice and digital connectivity within and outside of the fires cell.
- Coordinate with TACP.

Fire Support Specialist

4-63. Fire support specialists work under the supervision of the fire support operations noncommissioned officer. They support the operations and plans and the targeting element as directed. Their primary responsibilities follow—

- Operating the elements' computer systems.
- Supporting the development of fire support planning and targeting products as directed by the plans and targeting battle captain and targeting officers.
- Operating and maintaining voice communications equipment.
- Maintaining updated unit information— for example, fire support teams, combat observation and lasing teams, radar, battery, and mortar locations and digital/voice status.
- Maintain the current no-fire area list.
- Fire mission processing.
- Coordinating clearance of fires with adjacent and other affected units/assets.
- Operating and maintaining voice communications equipment.
- Operating assigned vehicles.

COLT

4-64. A *combat observation and lasing team* is a fire support team controlled at the brigade level that is capable of target acquisition and has both laser-range finding and laser-designating capabilities. It is designed to maximize the use of laser-guided munitions. COLT is organic to the BCT headquarters and headquarters company. Each BCT has five COLTs. The brigade FSO is responsible for training the COLT and for performing pre-combat checks and mission briefings/rehearsals before employment. Their employment is planned and executed under the supervision of the BCT fires cell. The COLT could be used as independent observers to weight key or vulnerable areas. Although originally conceived to interface with the copperhead, a COLT can be used with any munitions that require reflected laser energy for final ballistic guidance. The COLT self-location and target ranging capabilities allows first-round fire for effect with conventional munitions to be achieved.

4-65. The COLT gives the maneuver commander a powerful capability to attack point targets as well as area targets with accuracy. To maximize the effectiveness of the COLT, positioning must be carefully considered. To provide the best coverage and to allow the greatest survivability for the COLT, consideration should also be given to employing them as pairs, or with a company/troop fire support team, or partnered with elements of the reconnaissance squadron.

4-66. The brigade FSO or representative develops the observation plan by positioning COLT, joint fires observers, and joint terminal attack controllers to support the BCT commander's overall intent. The commander approves their positioning during development of the initial fire support plan, ISR plan, and airspace control plan. Joint terminal attack controllers or forward air controllers (airborne) may employ personnel conducting terminal guidance operations to facilitate close air support using terminal attack control procedures.

4-67. COLT provides observation of and can attack key targets as part of BCT operations. Support and security for them is a consideration, since they are extremely vulnerable if positioned forward of the maneuver battalions or reconnaissance squadron without security. Planning and integrating them into the reconnaissance squadron scheme of operations should always be considered, which will provide them some degree of security. COLT should always be an integral part of the BCT observation plan. They should not be positioned outside the range of friendly artillery during defensive operations. During offensive operations, they should not be positioned outside the ability of the BCT to ensure their support, protection/extraction, and the ability to communicate with the BCT tactical CP or main CP.

Tactical Air Control Party

4-68. An Air Force TACP (which is typically attached to an army CP) is collocated with the fires cell at the BCT main CP. The overarching mission of the BCT TACP is to plan, coordinate, and direct air support for land forces. The air component air liaison officer advises the BCT commander and staff on air support for BCT operations. The air liaison officer leverages the expertise of the TACP with linkages to the division and corps TACP to plan, to coordinate, to synchronize, and to execute air support operations. The TACP maintains situational understanding of the total air support picture.

4-69. The TACP attached to the BCT is sufficiently resourced to support BCT operations from both the tactical and main CP. The BCT TAC TACP focuses on execution of the current fight, immediate joint tactical air support requests, and current airspace requirements. The main TACP monitors current operations, and focuses on future operations, preplanned joint tactical air support requests, and future airspace requirements. The battalion level TACP includes an air liaison officer and joint terminal attack controllers. Air Force joint terminal air controllers are required at each maneuver battalion company and reconnaissance troop and are employed by the company/troop commander to provide close air support to support the company/troop. TACP coordinates activities through an Air Force air request net and the advanced airlift notification net. TACP functions include the following—

- Serving as the air component representative, providing advice to the BCT commander and staff on the capabilities, limitations, and employment of air support, airlift, and reconnaissance.
- Providing an air component coordination interface with not only the BCT fires cell, but also those of the battalions and squadron fires cells, and the air defense and airspace management/brigade aviation element.
- Assisting in the synchronization of air and surface fires and preparing the air support plan.
- Providing direct liaison for local air defense and airspace management activities.
- Integrating into the staff to facilitate planning air support for future operations and advising on the development and evaluation of close air support, air interdiction, reconnaissance, and J-SEAD programs.
- Providing terminal control for close air support and operating the Air Force air request net.

Other Joint and Army Augmentation to the Fires Cell

4-70. Joint and Army augmentation is essential to BCT operations. In addition to the Air Force TACP, other joint augmentation includes liaison officers to plan and coordinate naval surface fire support and Marine Corps support.

Naval Surface Fire Support Liaison Officer

4-71. The naval surface fire support liaison officer supervises a naval surface fire support team that may be attached to the BCT fires cell and coordinates and controls naval surface fires. The BCT naval surface fire support team communicates on the division/corps/theater army naval surface fire support high frequency net to gain naval surface fires. This net is also used for daily planning between the BCT and division/corps/theater army. Below brigade, BCT fire support digital and frequency modulation radio nets are used to exchange requests for naval fire support. When naval fire support is available and the general tactical situation permits its use, naval firepower can provide large volumes of devastating, immediately available, and responsive fire support to combat forces operating near coastal waters. These fires may be in support of amphibious operations within range of Navy aircraft and gunfire, but they also may be made available to support land operations. Normally, the fires cell coordinates naval surface fire support. However, a Marine Corps liaison element within the Marine Expeditionary Force headquarters provides task-organized, trained, and equipped teams to facilitate the planning, coordination, and terminal control of air, artillery, and naval surface fires when operating with multinational forces.

United States Marine Corps Liaison Officer

4-72. A Marine Corps (may also be a liaison team) may augment the fires cell based on METT-TC to coordinate naval and/or Marine Corps air support to the BCT. The fires cell processes requests for Navy/Marine Corps air support through this liaison and/or team. A firepower control team may be attached to the maneuver battalions and/or reconnaissance squadron to perform terminal control of Navy/Marine Corps air support. In the absence of an observer from the firepower control team, the company/troop fire support team or the Air Force joint terminal attack controller may control naval and/or Marine Corps air.

Nonlethal Fires Element (Main CP)

4-73. As described earlier, *nonlethal fires* are any weapon system's fires that do not directly seek the physical destruction of the intended target and are designed to impair, disrupt, or delay the performance of enemy forces, functions, or facilities. The fires cell's nonlethal fires element includes civil affairs, EW, and information engagement personnel. The brigade legal section is also collocated with the fires cell. Working under the guidance and direction of the executive officer, the nonlethal fires element with other staff cells, plans, coordinates, integrates, and synchronizes scalable fires to support BCT operations. This includes developing nonlethal fires related MOP and MOE.

Information Engagement Staff Offiecr

4-74. The S-7 is responsible for the planning, the coordination, the integration, and the synchronization of inform and influence activities for the BCT. Primary responsibilities include—

- Advising the BCT commander and staff on all aspects of inform and influence activities.
- Coordinating aspects of inform and influence activities with the fires cell nonlethal fires element.
- Ensuring that inform and influence activities are integrated into staff processes and orders.
- Maintaining friendly information situational awareness and providing relevant information to the BCT common operational picture.
- Conducting assessment of Army information tasks in BCT operations.
- Briefing deception operations. (See FM 3-13 for more details.)

Electronic Attack Officer

4-75. The EW officer provides the necessary EW subject matter expertise to support targeting, execution, and assessment for all BCT EW operations. Responsibilities include—

- Requesting and obtaining intelligence reports and identifying enemy intelligence targets.
- Recommending electronic attack objectives in developing TSS and high priority targets.
- Nominating targets for electronic attacks.
- Developing electronic attacks related to MOP and MOE for BCT assessment.
- Coordinating with the reconnaissance squadron for electronic attack operations to disrupt enemy or adversary.
- Recommending electronic attack employments for inclusion into the AGM, TSS, targeting synchronization matrix and fire support tasks.
- Recommending electronic attack objectives and synchronize reconnaissance squadron operations.
- Identifying potential conflicts of electromagnetic spectrum use by EW assets and coordinating deconfliction.
- Recommending electronic protect and EW support operations to support targeting.
- Deciding what EW tasks are essential to the success of future operations.
- Focusing where assets are deployed to detect HPT.
- Deciding whether the intended effect achieved by electronic attack was successful or not.
- Addressing who and when portion of task. (See FM 3-36 for more details on EW integration into the targeting process.)

Chapter 4

Psychological Operations Noncommissioned Officer

4-76. The psychological operations noncommissioned officer provides the necessary subject matter expertise support to targeting, execution, and assessment for all BCT military information support operations. Responsibilities include—
- Developing and recommending supporting military information support operations objectives and potential targets to the BCT commander.
- Writing the Appendix 10 - Information Engagement of ANNEX C – OPERATIONS of the BCT OPORD.
- Serving as the nonlethal fires noncommissioned officer in charge.
- Developing, related MOP and MOE, and monitoring the effectiveness of military information support operations for BCT assessment.
- Coordinating with public affairs regarding counterpropaganda efforts.
- Establishing voice and digital linkage with supporting military information support operations elements.
- Coordinating resources for supporting military information support operations elements.

Civil Affairs Officer

4-77. The BCT S-9 provides civil affairs expertise for the planning, coordinating, and the monitoring of civil affairs operations in the BCT area of operations. This by definition includes populace and resource control (including noncombatant evacuation operations and dislocated civilian operations), foreign humanitarian assistance, civil information management, nation assistance, and support to civil administration. Major functions include—
- Serving as the staff proponent for the organization, use, and integration of attached civil affairs forces.
- Developing plans, policies, and programs to further the relationship between the BCT and the civil component in the BCT area of operations.
- Serving as the primary advisor to the BCT commander on the effect of civilian populations on BCT operations.
- Assisting in the development of the plans, the policies, and the programs that are needed to deconflict civilian activities with military operations within the BCT area of operations. This includes displaced civilian operations, but is not limited to populace and resources control.
- Advising the BCT commander on legal and moral obligations incurred from the long- and short-term effects (economic, environmental, and health) of BCT operations on civilian populations.
- Coordinating, synchronizing, and integrating civil-military plans, programs, and policies with operational objectives.
- Advising on prioritizing and monitoring expenditures of allocated funds dedicated to civil affairs, operations, and facilitating movement, security, and control of funds to subordinate units. Coordinating with funds controlling authority, comptroller, or resource managers to meet the commander's objectives.
- Conducting, coordinating, and integrating deliberate planning for civil affairs operations in support of BCT operations.
- Coordinating and integrating area assessments and area studies in support of civil affairs operations.
- Advising the BCT commander and staff on protection of culturally significant property and facilities (religious building, shrines and consecrated places, museums, monuments, art, archives and libraries).
- Facilitating integration of civil affairs inputs to the BCT common operational picture.
- Advising the BCT commander on using military units and assets that can perform civil affairs missions.

4-78. Major capabilities of the BCT civil affairs staff include—
- Providing tactical level planning, management, coordination, and synchronization of key civil affairs functions within the BCT commander's area of operations.
- Providing a mechanism for civil-military coordination, collaboration, and communication within the BCT area of operations.
- Assisting the logistics officer with identifying and coordinating for facilities, supplies, and other material resources available from the local civil sector to support BCT operations.

4-79. The S-9 brings civilian considerations to the forefront during the targeting process to achieve the desired nonlethal effects on the host nation population. These actions help ensure that civilians have minimal impact on BCT tactical operations. The position is responsible for the following—
- Integrating civil affairs objectives/HPT with the BCT targeting process.
- Developing civil affairs related MOP and MOE for BCT assessment.
- Writing Annex K, the civil affairs operations annex to the BCT OPORD.
- Conducting liaison with key civilian authorities and leaders in the BCT area of operations.
- Synchronizing civilian relief effort with BCT objectives.
- Providing a direct linkage with the civil-military operations center (when established).
- Providing the BCT commander, staff and subordinate elements with regional/cultural expertise through reachback.

Note: See FM 3-05.40 and FM 3-05.401 for more information on civil affairs.

Public Affairs Officer

4-80. The BCT public affairs officer roles include serving as the principle adviser to the commander and staff for media engagement and conducting media operations. Public affairs officer has a legal responsibility to factually and accurately inform various publics—domestic and foreign—without intent to propagandize or change behavior. The public affairs officer plans and executes Soldier and community outreach both foreign and domestic. The public affairs officer and staff provide training and support to stability operations in coordination with the S-9 and civil affairs staff in the fires cell nonlethal fires element.

Public Affairs Noncommissioned Officer

4-81. The public affairs noncommissioned officer supervises day to day operation of public affairs for the BCT, directs activities of the public affairs specialist and public affairs broadcast specialist, and leads the public affairs activities of the BCT in the absence of the public affairs officer.

Public Affairs Specialist/Public Affairs Broadcast Specialist

4-82. The public affairs specialist/public affairs broadcast specialist works under the supervision of the public affairs noncommissioned officer to support BCT public affairs activities as directed. The public affairs specialists have the joint responsibilities which include the development of public affairs planning and targeting products (working with the fires cell battle captain and targeting officers) and, as directed by the public affairs officer and public affairs noncommissioned officer operating and maintaining voice communications equipment, coordinating media escorts and operating assigned vehicles. They also have the following specific responsibilities—
- **Public Affairs Specialist**. The public affairs specialist coordinates release of print and Web-based command information products to higher echelon headquarters, responds to media requests/queries, supports command information programs, coordinates the Hometown News Release Program, acquires and transmits photographs and information strategy products between higher and lower echelon, and performs limited newsgathering.
- **Public Affairs Broadcast Specialist**. The public affairs broadcast specialist coordinates release of radio and television and visual information products to higher echelon headquarters; acquires and transmits audio-visual, radio/television and Web-based products for higher and lower

echelon headquarters, family members, and unit support groups; coordinates the radio and television hometown news release program; performs limited electronic audio and video newsgathering; and determines visual information requirements.

Note. See FM 3-61.1 for more information on public affairs.

Brigade Legal Section

4-83. The brigade judge advocate, along with the trial counsel and the paralegal noncommissioned officer, forms the brigade operational law team. The brigade operational law team is a section positioned with the main CP as part of nonlethal fires element. The section is deployable forward in whole or in part as directed by the brigade judge advocate. The brigade judge advocate serves both as a personal staff officer to the BCT commander and a special staff officer. The legal team provides legal advice during the MDMP and all other planning and targeting working group sessions conducted by the BCT staff. The members of the brigade legal section serve as subject matter experts on rules of engagement and rules of interaction, targeting, international law, law of war (including treatment of detainees, enemy prisoners of war, civilians on the battlefield, and other noncombatants) and all other legal aspects of BCT operations. In addition to providing support in international and operational law, the brigade legal section provides advice and support to the commander in administrative and civil law, contract and fiscal law, military justice, claims, and legal assistance. The paralegal noncommissioned officer provides administrative and paralegal support to the judge advocates in the brigade legal section and supervises the paralegals in BCT battalions.

Note. See FM 1-04 for more information on legal support to the targeting process.

Tactical Command Post Fires Element

4-84. The designated members of the tactical CP fires element are part of the fires cell's lethal and nonlethal fires elements in the BCT main CP, or are deployed forward with the tactical CP when it is deployed. When deployed with the tactical CP the fires element tracks and maintains situational understanding of all fire support assets. Its main function is to execute current operations, focusing on the decisive fight. Functions of the tactical CP fires element include the following—

- Monitoring the tactical situation.
- Maintaining and updating unit information and digital/voice status.
- Ensuring tactical fire control with supporting field artillery and target acquisition assets.
- Monitoring processing of preplanned fires in the fire support plan.
- Coordinating clearance of all fires with units.
- Maintaining and updating the current active no fire area list.
- Maintaining digital link to field artillery and target acquisition assets.
- Tracking and maintaining situational understanding of close air support.
- Tracking and maintaining situational understanding of naval surface fire support.
- Sending fire missions to the battalion fire direction center for processing.
- Requesting assessment reports.
- Ensuring mission fired reports and artillery target intelligence reports are received and processed.

FIRE SUPPORT COORDINATION ORGANIZATIONS AT BCT SUBORDINATE ECHELONS

4-85. The BCT has organic fire support organizations: battalion/squadron fires cells and company/troop fire support teams that work closely with the battalion/squadron fires cell. These organizations are vital parts of the combined arms infrastructure that exists within the BCT.

4-86. Fire support organizations in the maneuver battalions and the reconnaissance squadron support their respective commanders but work closely with the BCT fires cell. The brigade FSO will advise the BCT commander on training, personnel management, maintenance, and equipment readiness for all subordinate fire support organizations. The maneuver battalions and the reconnaissance are each organized with a fires cell and each have an Air Force TACP. The fire support teams are assigned to the battalion/squadron headquarters and headquarters companies to facilitate training, but deploy to the maneuver companies and the reconnaissance troops for tactical operations. The infantry company fire support team includes platoon forward observers for each of its platoons. The brigade troops battalion has fire support planners in its S-3 section to help the commander and staff plan and execute sustainment area fires up to a Level II threat.

SUBORDINATE FIRES CELLS AND AIR COMPONENT TACP

Battalion/Squadron Fires Cell

4-87. The battalion/squadron fires cell provides an organic fire support coordination capability within the reconnaissance squadron, brigade special troops battalion, and maneuver battalion headquarters. The fires cell assists the battalion/squadron in executing its portion of the BCT scheme/concept of fires as well as their own scheme/concept of fires. Through its computer systems, the fires cell provides the company/troop fire support team digital linkage to the battalion/squadron mortars as well as to fire support assets available at the BCT or higher levels.

Fires Support Officer

4-88. The FSO is a field artillery officer. The FSO serves maneuver units at all levels from company up to theater army. FSO is responsible for either advising the force commander or assisting the chief of fires or FSCOORD to advise the force commander on fire support matters. The battalion/squadron FSO is responsible for the planning, the coordination, and the execution of fire support for the battalion/squadron commander's concept of operation. FSO responsibilities include the following—

- Advising the commander and his staff on fire support matters. This includes making recommendations for integrating battalion/squadron mortars into the scheme/concept of fires and their movement in the scheme of maneuver.
- Supervising all functions of the battalion/squadron fires cell.
- Ensuring all fire support personnel are properly trained to support battalion/squadron operations.
- Preparing and disseminating the fire support execution matrix and/or the fire support plan.
- Assisting in the coordination for positioning or movement of fire support assets in the battalion/squadron area of operations.
- Conducting bottom up refinement of the BCT fire support plan.
- Directing development of battalion/squadron fire support tasks.
- Coordinating with the TACP on close air support missions and for terminal control personnel.
- Providing coordination channels to the BCT fires cell nonlethal fires element for inform and influence activities or other scalable fires related support.
- Planning, directing, and monitoring the employment of laser designators where they will best support the commander's concept of operation.
- Translating the commander's intent into attack guidance for orders.
- Establishing and maintaining communications with the BCT fires cell, subordinate unit fire support teams, and the battalion/squadron mortars.
- Participating in fire support rehearsals.
- Processing requests for additional fire support with the BCT fires cell.
- Providing staff supervision of the field artillery assets attached or under the operational control of the battalion/squadron.
- Disseminating the approved target list and execution matrix to subordinate elements.
- Recommending appropriate changes in the target list and attack guidance when required.

Assistant Fire Support Officer

4-89. The battalion/squadron assistant FSO acts as the battalion/squadron FSO in the fire support officer's absence. The assistant FSO interfaces with the battalion/squadron S-2 and provides the S-2 and the

Chapter 4

battalion/squadron FSO with information on the vulnerabilities of targets. The vulnerabilities of targets consist of specific requirements for accuracy of target location assurance, level of target description, and duration the target may be considered viable for attack by fire support systems. The responsibilities of the assistant FSO include—

- Helping the battalion/squadron S-2 write their target acquisition and ISR plans.
- Helping provide staff supervision of the target acquisition assets attached, organic, and under the operational control of the battalion/squadron.
- Developing, recommend to the commander, and disseminating the AGM to subordinate elements; recommending changes in attack guidance.
- Determining, recommending, and processing time-sensitive HPT to the BCT fires cell.
- Coordinating with the battalion/squadron S-2 for target acquisition coverage and processing of battalion/squadron HPT.
- Producing a TSS matrix for target acquisition assets with the battalion/squadron S-2.
- Coordinating integration of fire support into battalion/squadron tactical operations center operations, to include physical arrangements of fire support equipment, responsibilities involving tactical operations center operations, and security.

Battalion Targeting Officer

4-90. The targeting officer supervises the counterfire operations section and the target processing element within the battalion CP. The targeting officer serves as the battalion counterfire officer. The responsibilities of the targeting officer include—

- Advising the commander on acquisition system employment, capabilities, and limitations.
- Performing predictive analysis of enemy fire support locations.
- Assisting in the integration of intelligence and other war fighting functions.
- Assisting the intelligence war fighting function with the integration of ISR collection assets.
- Assisting in target production by developing the HPTL, AGM, and TSS.
- Directing the targeting meeting.

Target Acquisition Platoon Leader

4-91. The target acquisition platoon leader (131A Warrant Officer) supervises the battalion's target acquisition platoon and serves as the brigade assistant counterfire officer when positioned with the BCT fires cell. The responsibilities of the target acquisition platoon leader include—

- Performing tactical coordination for field artillery radars, survey and met in support of the supported higher headquarters, to include communications, security, protection, logistics, and administration.
- Advising the commander on acquisition system employment, capabilities, and limitations.

Fire Support/Targeting Noncommissioned Officers

4-92. The fire support and targeting noncommissioned officers together with the assistant FSO and fire support sergeant provide a 24-hour capability to plan and coordinate targeting operations. Their primary responsibilities include the following—

- Coordinating the close air support operations with TACP representative.
- Operating and maintaining the targeting computer systems.
- Maintaining the targeting common operational picture display.
- Maintaining the target production display.
- Updating and purging targeting files as directed by the BCT targeting officer.
- Ensuring targets that are acquired are processed to the appropriate fire support assets in accordance with the targeting synchronization matrix.
- Ensuring essential voice and digital connectivity within and outside of the fires cell.

Fire Support Sergeant

4-93. The battalion/squadron fire support sergeant is the enlisted assistant to both the battalion/squadron FSO and assistant FSO. His responsibilities include the following—
- Training and validating enlisted personnel of the battalion/squadron fires cell and fire support team.
- Assisting the battalion/squadron FSO in developing fire support tasks to support battalion/squadron operations.
- Ensuring voice and digital connectivity with the BCT fires cell, supported and supporting units, and fire support assets.
- Planning and coordinating all support (administrative and logistical) for the fires cell.
- Maintaining and updating the fire support status charts and situation maps.

Fire Support Specialist

4-94. The responsibilities of the fire support specialist include—
- Operating and maintaining the fires cell's equipment, including computer equipment.
- Helping in fire support planning and coordination.
- Operating and maintaining communications equipment.
- Preparing and maintaining a situation map.
- Preparing and posting daily staff journals and reports.
- Assisting in establishing, operating, and displacing the fire support equipment.

Air Force Tactical Air Control Party

4-95. The Air Force liaison element to the battalion is identified as the TACP. The TACP primary mission is to advise the battalion commander on the capabilities and limitations of air power and assist the ground effort in planning, requesting, and coordinating close air support. The TACP/joint terminal attack controller is capable and authorized to perform terminal attack control of close air support for the battalion.

Note. See ATTP 3-09.36 for capabilities and tasks that are unique and significantly challenged by close air support at the tactical level.

4-96. Joint terminal attack controllers in the Air Force TACP—
- Coordinate with the joint fires observers.
- Know the enemy situation, selected targets, and location of friendly units.
- Know the battalion/squadron plans, position, and needs.
- Validate targets of opportunity in regard to the target's proximity to friendly elements and that targets are accurately located.
- Advise the commander on proper employment of air assets.
- Submit immediate requests for close air support.
- Control close air support with supported commander's approval.
- Perform BDA.

PREPARING AND CONDUCTING TARGETING WORKING GROUP

PREPARING FOR TARGETING WORKING GROUP SESSIONS

4-97. Preparation and focus are keys to success of the BCT targeting working group. The working group performs targeting functions and represents the interest of the commander. These include—
- Assessing previous executed targeting cycles.
- Providing relevant information and analysis.
- Maintaining running estimates and making recommendations.

- Preparing targeting products.
- Monitoring operations.
- Assessing the progress of operations.
- Consolidating proposed/draft decision points.
- Consolidating draft commander's critical information requirements for each target.
- Integrating necessary enablers into the concept for each target.
- Preparing input to targeting fragmentary order.
- Updating targeting synchronization matrix.
- Updating cover pages of targeting packet (baseball cards).

4-98. Each representative must come to each session prepared to discuss available assets, capabilities, limitations, and BDA requirements related to fires, intelligence, and maneuver functions. This means participants must conduct detailed prior coordination and be prepared to conduct planning, coordination, and deconfliction associated with targeting and operations. This preparation must be focused around the BCT commander's intent and a solid understanding of the current situation.

4-99. The BCT S-3 must be prepared to provide the following information—
- Current friendly situation.
- Maneuver assets available.
- Current combat power.
- Requirements from higher headquarters (including recent fragmentary orders or tasking).
- Changes to the commander's intent.
- Changes to existing fire support tasks.
- Changes to the task organization.
- Planned operations.

4-100. The BCT S-2 must be prepared to provide the following—
- Current enemy situation.
- Current ISR plan.
- Planned enemy courses of action (situation template) tailored to the time period discussed.
- ISR and target acquisition collection assets available and those the S-2 must request from higher headquarters.

4-101. The brigade FSO must be prepared to provide the following—
- Fire support assets/resources available.
- Proposed HPTL, TSS, AGM, targeting synchronization matrix, and changes to fire support tasks for the time period discussed.
- Recommended changes to FSCM for the period being discussed.

4-102. The specific situation dictates the extent of the remaining targeting working group member's preparation. They prepare to discuss in detail (within their own warfighting functional or staff section area) available assets and capabilities, the integration of their assets into targeting decisions, and the capabilities and limitations of enemy assets. The following tools should be available to facilitate the conduct of the targeting working group: HPTL, TSS, AGM, consolidated matrix to include the targeting synchronization matrix, fire support task(s) or other product(s) per standing operating procedure, a list of delivery assets/resources, and a list of collection assets/resources. For instance—
- **Targeting synchronization matrix.** The targeting synchronization matrix visually illustrates the HPT and designed to list specific targets with locations, in each category. The matrix has entries to identify if a target is covered by a named area of interest; the specific detect, deliver, and assess assets for each target; and attack guidance. Once completed, the targeting synchronization matrix serves as a basis for updating the ISR plan, observation plan, unit airspace plan, and issuing a fragmentary order once the targeting working group concludes its session. In addition, it facilitates the distribution of results produced by the targeting working group.

- **List of potential detection and delivery assets/resources.** A list of all potential detection and delivery assets/resources available to the BCT helps all attendees visualize what assets may be available for detection and delivery purposes. It is essential that staff members are prepared to discuss the potential contribution for the particular assets within their area of expertise, and able to identify terrain, airspace, and frequency spectrum requirements to ensure assets are a feasible solution.

CONDUCTING THE TARGETING WORKING GROUP SESSION

Targeting Working Group Typical Agenda

4-103. The brigade FSO chairs the targeting working group and is responsible for keeping it focused. The FSO opens each session of the targeting working group by conducting a roll call, and then briefly explains its purpose. This position prepares, and describes the agenda and specifies the time period to be addressed. The FSO is the arbitrator of any disagreements that may arise and ensures the session stays on track with the stated purpose and consistent with the BCT commander's guidance and intent. The brigade FSO actions empower the targeting working group to make adjustment within their area of expertise.

4-104. The following agenda (table 4-1) provides information covered by core targeting working group members. The agenda helps to validate the targeting working group visualization. The visualization allows the commanders to develop their intent and planning guidance for the operations.

Table 4-1. Targeting working group agenda (example)

AGENDA	
WHO	*WHAT*
S-2 Staff Representative	•Weather •Enemy situation and decision points (event template) •Battle damage assessment for targets engaged since last session •Analysis of enemy most likely and dangerous courses of action for next 24-72 hrs •Recommended changes to priority intelligence requirements •Intelligence, surveillance, and reconnaissance plan
S-3 Staff Representative	•New requirements from higher headquarters •Summarizes current situation •Provides status of combat power •Commander's guidance and intent •Planned operations during the focus period
Brigade Fire Support Officer or Fires Cell Representative	•Briefs current targeting products including the high-payoff target list, attack guidance matrix, target selection standards, targeting synchronization matrix, and fire support tasks •Status of fire support assets •Approved preplanned air requests and targets planned for next two days air cycles •Proposed high-payoff target list with target locations for concurrence and approval •Recommend, in conjunction with the Air Force air liaison officer, changes to working preplanned air requests
Air Liaison Officer	•Advises the group on the capabilities and limitations of air power
Fires Cell Targeting Officer	•Briefs high-payoff targets that have been attacked and associated battle damage assessment •Provides radar status and active radar zones •Briefs counterfire situation
Air Defense and Airspace Management/Brigade Aviation Element	•Identify airspace requirements •Receive airspace control measure requests •Identify airspace conflicts •Submit approved airspace control measure requests to higher
Information Engagement Officer	•Provides assessment of inform and influence type targets •Recommends new inform and influence type targets
Electronic Warfare Officer	•Brigade combat team electronic warfare plan •Deconflict frequency utilization with the signal staff officer
Others as required	•Judge advocate, psychological operations noncommissioned officer, civil affairs staff officer, liaison officers, and others provides amplification as required

Staff Participation

4-105. Maximum participation by the staff is essential. Staff members and warfighting function representatives must share their expertise and respective running estimate information on the capabilities and limitations of both friendly and enemy systems. They should also consider providing redundant means, if feasible, to decide, detect, deliver, and assess (D3A) targets.

Agenda

4-106. After the brigade FSO opens the session, the S-2 provides an intelligence update. The S-2 briefs the current enemy situational using event templates with current HVT locations, the commander's critical information requirements, named areas of interest, and conduct an overview of the current ISR plan. The S-2 continues with the BDA on targets previously engaged since the last session of the targeting working group and the impact on the enemy COA. Most importantly, the intelligence section prepares a predictive analysis of future enemy COA for the next 24–72 hours using the event template and a list of HVT. Finally, the agenda provides for an opportunity to brief changes to the commander's critical information requirement. The intelligence section products must be tailored to the designated time period to be discussed at the session which generally includes—

- The enemy situation.
- Review of the current ISR plans.
- BDA of targets engaged since the last session of the targeting working group and the impact on the enemy COA.
- An analysis of the enemy's most probable COA and locations for the next 24 to 36 hours (possibly projecting out 72 hours for targets subject to air attacks).
- Recommended changes to the PIR for the commander's approval when present, or review by the staff.

4-107. The S-3 discusses any particular guidance from the commander, changes to the commander's intent, and any changes since the last session of the targeting working group to include task organization, requirements from higher headquarters to include recent fragmentary orders and taskings, current combat power, the current situation of subordinate units, planned operations, and maneuver assets/resources available. Finally, the S-3 informs the staff of the status of assets/resources available for the targeting process. The operations section products must be tailored to the designated time period to be discussed at the session but generally include a friendly situation update that—

- Briefs any new requirements from higher headquarters since the last targeting working group session.
- Summarizes the current tactical situation.
- Informs on the status of available assets/resources (combat power).
- If the commander is not present, briefs any particular guidance from the commander and changes to his intent.
- Briefs planned operations during the period covered by the targeting working group session.

4-108. The brigade FSO briefs fire support assets available including status of fire support tasks, radars, close air support sorties available, status of naval surface fire support, and ammunition availability, HPTL, TSS, AGM, and targeting synchronization matrix.

4-109. The brigade FSO reviews the approved preplanned air requests for the period and those planned for the next two joint air tasking cycles (this may be briefed by the Air Force air liaison officer)—normally done in 24 hour increments. The Air Force air liaison officer advises and assists the FSO who recommends changes to the preplanned air support requests and target nominations for the planning cycle. The position provides proposed targeting guidance for the designated periods, a new targeting synchronization matrix with the proposed list of HPT, and locations for the staffs' concurrence and refinement. Once changes have been made to the HPTL. The brigade FSO facilitates a crosswalk to complete the rest of the matrix by identifying a detector, determining an attack means, and assigning an asset to assess each HPT—

- **Step One**. The first step is to select, or update the HPTL. These targets are derived from the S-2's list of HVT.

- **Step Two**. The next step is to determine and prioritize collection assets responsible for detecting, confirming, or denying the location of each suspected target or HPT. This information should then be entered into the detect portion of the targeting synchronization matrix. Be specific, state what unit or asset must detect, confirm, or deny the location of each specific target. Clear and concise tasking is given to the acquisition assets/resources. Mobile HPT is detected and tracked to maintain current target location. Task external air assets with a joint tasking air support request and record the joint tasking air support request number for tracking higher headquarters approval and processing during air tasking cycle. Submit airspace control means requests is for those air assets that require airspace use. Assets/resources are placed in the best position according to the running estimates of when and where the enemy targets will be located. Consider assigning a named area of interest to the target and enter the number on the targeting synchronization matrix.
- **Step Three**. The third step is to determine which attack asset/resource will be used to attack each target once detected or confirmed by using the list of delivery assets/resources available. Enter this information into the deliver portion of the targeting synchronization matrix. The effects of scalable fires are considered depending on the commander's targeting guidance and desired effects. Consider redundant means to attack each target. The attack guidance is entered when determining an attack asset/resource for each target. Determine for each delivery means when to attack the target (immediately, as acquired, or planned) and the desired effects for each target. For instance, the effects of fire support can be to deceive, degrade, delay, deny, destroy, disrupt, divert, exploit, interdict, neutralize, or suppress the target.
- **Step Four**. The final step is to determine and prioritize which assets will assess how well the attack was executed and if the attack resulted in the desired effects. Enter this information into the assess portion of the targeting synchronization matrix.

4-110. At the conclusion of the crosswalk, the targeting synchronization matrix should be complete. The brigade FSO should keep the focus of the discussion to within the possibilities of friendly unit operations and should be the final arbitrator when completing the targeting synchronization matrix.

4-111. Based on the situation, additional staff members will need to provide the capabilities and limitations of their available assets/resources. They must be prepared to discuss the integration of their assets/resources into the targeting process. Additionally, they must also be able to discuss in detail the capabilities and limitations of enemy assets within their area of expertise. If, it is impossible for a particular staff officer to attend the session, they must provide their products and information to the primary staff officer that has supervisory responsibility for their particular area.

SUBSEQUENT ACTIONS

4-112. Upon completion of the targeting working group session, the commander is briefed on the results. Once the results of the targeting working group session are approved targeting products are updated, written, and reproduced for distribution. This must be accomplished quickly, allowing sufficient time for subordinate units to react, plan, rehearse, and execute. Targeting working group products include—
- Update HPTL, TSS, AGM, and observer overlay. These, with data from the ISR plan may be combined into a unit specific targeting synchronization matrix.
- Updated fire support tasks.
- Update tasking(s) to subordinate units and assets. The S-3 should prepare and issue a fragmentary order to subordinate elements to execute the planned attack and assessment of targets developed by the targeting working group.
- Update ISR plan. The S-2 reorients his acquisition assets and updates and disseminates the ISR plan.
- Update unit airspace plan. The air defense and airspace management/brigade aviation element ensures all brigade airspace requirements for organic and supporting air assets are integrated at the BCT and passed to the airspace element.

4-113. After the targeting process is completed, the staff obtains the commander's approval and then prepares fragmentary orders with new tasks to subordinate units. The plan is rehearsed, if time permits. Targeting actions continue using the targeting products the unit has adopted.

SYNCHRONIZATION

4-114. The key to effective synchronization of targeting is the thorough use of the targeting process in the planning, preparation, and execution of the maneuver plan. As the commander and staff form the operations plan during the war gaming process, the decision support template is developed. It is the key to synchronizing the fire support plan with the scheme of maneuver. The decision support template facilitates the BCT commander's staff war gaming. It also identifies critical fire support triggers on the battlefield and is an aid in synchronizing the warfighting functions. The war gaming process identifies the decision points for the commander. The decision support template graphically portrays the decision points and the options available to the commander if an action occurs. The decision support template provides the information required to provide effective fires in support of the maneuver force.

4-115. The attack guidance provided fire support personnel and units must define how, when, with what restrictions, and in what priority to attack different targets. This should include guidance on the following—

- Final protective fires.
- Screening fires.
- Obscuration.
- Illumination.
- Positioning.
- Engagement method.
- Counterfire targets.

4-116. Firing units must know the critical time and location they must be in to support each phase of an operation.

4-117. The commander and brigade FSO remembers that all tasks must be assigned. If a task is not assigned to an individual, everyone will believe it is someone else's responsibility. Unassigned tasks may never be carried out. For example, simply assigning responsibility for firing on a planned target is not enough. The criteria for firing must be made clear. Previsions are made to ensure the forward observer or FSO in questioned is fully aware of his responsibilities and will be able to carry out the task.

4-118. The fire support execution matrix is the blueprint for executing the fires portion of the OPORD and should correspond to the synchronization matrix. There is no specific format for how a fire support execution matrix is set up. An example is shown below. The matrix is clear, simple, and concise. It communicates the commander's concept of fires and plan for execution. The synchronization matrix is easy to develop and detailed enough to implement. It should be tied to the events on the decision support template. The BCT fire support execution matrix allocates resources and assigns responsibilities for observing and executing BCT targets. (See table 4-2.) It defines the transition from the BCT to the task force fight. The task force fire support execution matrix is a stand-alone document. It is detailed enough for battalion FSO and company FSO to assume control and execute the task force commander's intent for fire support. See FM 6-20-40 and FM 6-20-50 for additional information.

Table 4-2. BCT fire support execution matrix (example)

Phase	Advance	Assault	Advance	Secure
POF / PRI TGT	2AR / 6	77IN	78AR	2AR
TF 77IN		AB 1000 AB 1004	AB 1010	
TF 1AR			AB 1009 AB 1008	

Table 4-2. BCT fire support execution matrix (example)

Phase	Advance	Assault	Advance	Secure
TF 78AR		AB 1002 AB 1003 AB 1005		A4B
TF 2AR	A1B	AB 1001 AB 1006 AB 1007		A3B
Bde	A2B *Bde PREP	CAS *Div PREP	CAS	CAS
CFL	PL Horse	O/O PL Dog		

Legend:
A#B – group of targets
CAS – close air support
O/O – on order
PRI – priority
AB #### - target number
CFL – coordinated fire line
PL – phase line
PRI TGT – priority target
AR – armor
Div - division
POF – priority of fire
TF – task force
Bde - brigade
IN – infantry
PREP – preparation

4-119. At task force level, the battalion FSO prepares the fire support execution matrix. (See table 4-3.) The battalion FSO coordinates with the company FSO and mortar platoon leader. In conjunction with the task force S-3, the battalion FSO positions and controls the organic mortars of the task force. With the S-2, the FSO positions and controls observation assets. This coordination is needed to ensure the fire support plan—

- Meets the commander's guidance.
- Avoids unplanned duplication.
- Uses all assets assigned to the task force.
- Assigns observers and backup observers for all task force targets and BCT targets assigned to the task force.
- Specifies who, when, where, and how for detecting and delivering fires on targets.

Table 4-3. Task force fire support execution matrix (example)

	AA	LD/LC	PL Gun	PL Pistol	PL Saber	
Team Tank	FA PRI AB 3002	FA POF FA PRI C3B	Series Finish	FA POF 155 mm FPF		
Team A	MORT B POF MORT B PR AB 3110	MORT B POF MORT B PR AB 3119	MORT POF	MORT B POF 155 mm FPF	MORT B POF	7
Team C	MORT A POF MORT A PR AB 3207					6
Team D		MORT A POF MORT A PR AB 3216			MORT A POF	5
Scouts	FA POF					4
TF	Group C4B Series Joe	ACA ORANGE TOT 0800	F16 (ground alert) 0815 - 1015	Groups C7B, C8B, C9B, ACA Grape 0/0	FA POF Groups C12B & C13B	3
Attack Aircraft		TAI 6		TAI 5		1
	A	B	C	D	E	

Legend:
AA – assembly area
FA – field artillery
MORT – mortar
PRI TGT – priority target
ACA – airspace coordination area
FPF – final protective fire
O/O – on order
TAI – target area of interest
A#B – group of targets
LC – line of contact
POF – priority of fire
TF – task force
AB #### - target number
LD – line of departure
PRI – priority
TOT – time on target

4-120. Clear, well practiced standard/standing operating procedures (SOP) within division, BCT, and task force teams are essential to synchronizing the targeting effort. The SOPs must specify when and how tasking and requests for support and information are to be passed.

Chapter 4

4-121. Rehearsals are required to build confidence and understanding among all warfighting functions before combat operations. Rehearsals at all levels are key to understanding—
- The concept of the operation.
- Verifying specific responsibilities and timing.
- Backup procedures to help synchronize unit operations.

4-122. A rehearsal is an effective tool for identifying and refining concept of operations. However, rehearsals should not be used for making major changes to the plan. Any last minute, major changes to the scheme of maneuver made during rehearsals may cause a reduction in the effectiveness of fire support.

4-123. The key fire support points that should be emphasized during rehearsals are—
- Positioning and movement plans of fire support and target acquisition systems are synchronized with the maneuver concept of operations and the ISR plan.
- Target acquisition plan under the supervision of the field artillery S-2 is verified.
- Integration of the target acquisition and intelligence collection plan is verified.
- Fire support plan incorporating scalable fires and electronic attacks are validated with the scheme of maneuver, the commander's intent, and the attack guidance.
- Obstacle and barrier plan of the maneuver force is incorporated into the fire support plan.
- Fire support coordination measures and the airspace coordinating measures are fully integrated with maneuver graphic control measures.
- Target locations, engagement criteria, rules of engagement, and trigger points or events to initiate attack are verified.
- Fire support plan and ISR plan are synchronized with the maneuver concept of operations and meets the commander's intent.
- Primary and backup observers for each target are assigned to support the fire support plan.
- Battlefield handoff points to indicate transition from the BCT fight to the battalion task force fight are clearly identified; for example, phase lines, terrain features, grid coordinates, and so forth.
- Responsibilities for clearance of fires are clearly spelled out.

4-124. Rehearsals are conducted early enough for essential personnel to attend, disseminate and implement minor changes, and get into position before plan execution. If time does not permit a complete rehearsal with all essential personnel and equipment, some form of rehearsal must take place with all key leaders. It can be as simple as a leader discussion over a sand table or a radio rehearsal tactical exercise without troops over similar terrain. Rehearsals provide the commander and the brigade/battalion FSO with a final opportunity to synchronize the fire support plan with the scheme of maneuver before the battle. If possible, the fire support plan should be rehearsed with the maneuver plan. A combined rehearsal improves responsiveness of fires and synchronization of all warfighting functions. At the end of an effective rehearsal, everyone involved in the detecting and delivery functions of the targeting process should know their responsibilities and the cues for action.

Appendix A
Find, Fix, Track, Target, Engage, and Assess

The dynamic targeting process occurs during the detect, deliver and assess functions of the decide, detect, deliver, and assess (D3A) methodology and phase 5: mission planning and execution of the joint targeting cycle. Decisions have already been made, orders have already been published, and subordinate units are determining how to execute the attacks, or may even be in the middle of execution.

Dynamic targeting is required because not all targets are located accurately enough or identified in sufficient time for inclusion in the deliberate targeting process. A target of opportunity may emerge, or a change in the situation may necessitate a change to a planned target. These targets still require confirmation, verification, validation, and authorization, but in a shorter timeframe than the deliberate targeting process allows.

Dynamic targeting is primarily designed to attack time-sensitive targets (TST) and high-payoff targets (HPT). Lower priority targets are normally not worth redirecting assets from previous planned or assigned missions.

The process developed to facilitate dynamic targeting at the joint level is find, fix, track, target, engage, and assess (F2T2EA). While the steps are listed in the order presented to ease explanation, several steps are accomplished simultaneously and overlapped. For example, the track step frequently continues through the completion of the assess step.

STEP 1 – FIND

A-1. The find step involves intelligence collection based on joint intelligence preparation of the operational environment (JIPOE). Traditional intelligence, surveillance, and reconnaissance (ISR) assets, as well as nontraditional assets, may provide initial detection of an entity.

A-2. Each entity is immediately evaluated as a potential target. Based on the situation and the commander's guidance, some entities are clearly identified not a target. Other entities may be clearly identified as a target already included in the targeting process. The remaining entities display some characteristics of a target, but need more analysis to categorize them properly. These entities requiring further analysis are called emerging targets.

EMERGING TARGET

A-3. The term "emerging target" is used to describe a detection that meets sufficient criteria to be developed as a potential target using dynamic targeting. The criticality and time sensitivity of an emerging target, and its probability of being a potential target, is initially undetermined. Emerging targets normally require further reconnaissance and/or analysis to develop, confirm, and continue dynamic targeting.

A-4. During the find step, an emerging target will be—
- Designated a probable target or identified as a TST and the dynamic targeting process is continued.
- Designated a probable target not requiring dynamic targeting and passed to deliberate targeting.
- Discarded completely or entered on the no-strike list.
- Analyzed until a determination can be made (that is, continuing the find step).

Appendix A

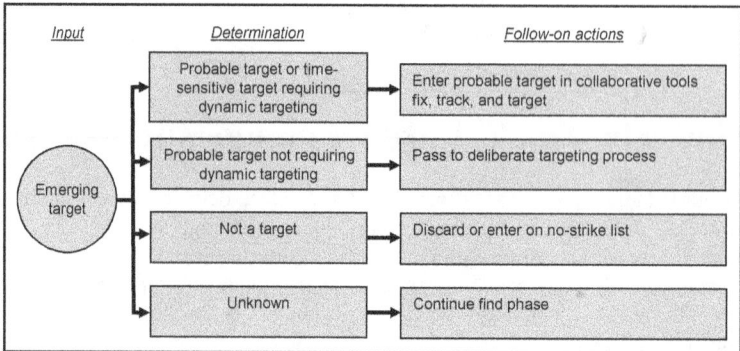

Figure A-1. Find step determinations and follow on actions

A-5. Sometimes the entire dynamic targeting process can occur within the span of a few minutes. An aircraft returning from a mission detects and identifies an emerging target, and determines it to be a potential target. The aircraft commander relays the information to higher and receives approval to engage the target.

A-6. In this example, find and fix steps are completed nearly simultaneously without the need for traditional reconnaissance. The aircraft commander continues to track the target during an abbreviated coordination and approval process. The entity is targeted, engaged, and an initial assessment is conducted by the same system that initially detected the target.

INPUTS

A-7. Inputs to the find step—
- Clearly delineated joint force commander (JFC) dynamic targeting guidance and priorities.
- Focused JIPOE to include identified named areas of interest, target areas of interest, and cross cueing of intelligence disciplines to identify potential target deployment sites or operational environments. (The Army still uses the IPB process.)
- Collection plans based on the joint intelligence preparation of the operational environment.

OUTPUTS

A-8. Outputs of the find step—
- Potential targets detected and nominated for further development within dynamic targeting.

STEP 2 - FIX

A-9. The fix step of dynamic targeting includes actions to determine the location of the probable target. This step also results in a positive identification of a probable target as worthy of engagement as well as determining its position and other date with sufficient fidelity to permit engagement.

A-10. The fix step begins after a probable target requiring dynamic targeting is detected. When a probable target is identified, sensors are focused to confirm target identification and precise location. This may require diverting assets from other uses. The collection manager may have to make a decision on whether the diversion of reconnaissance assets from the established collection plan is merited. The collection, correlation, and fusing of data continues in order to confirm that the target meets the criteria to be classified as a TST or other target requiring dynamic targeting.

A-11. Completing the fix step in a timely manner requires reconnaissance and surveillance with the capability to identify stationary and mobile targets, day or night, in any weather conditions, through all forms of terrain, camouflage, or concealment, to the degree of accuracy required by the engaging weapon systems. The ISR assets must also provide both operators and intelligence analysts with the location of the target with an accuracy that allows engagement by available weapons systems.

A-12. An unanticipated or unplanned TST may be identified during the fix step, requiring JFC confirmation and classification as a TST. The determination or estimation of the target's window of vulnerability defines the timeliness required for successful prosecution, and influences the required prioritization of assets and the risk assessment. TST receive the highest priority in dynamic targeting.

INPUTS

A-13. Inputs to the fix step—
- Probable targets requiring dynamic targeting.
- Sensor information on the target.

OUTPUTS

A-14. Outputs of the fix step—
- Target identification, classification, and confirmation.
- Target location accuracy refined to the level required for target engagement.
- Determination or estimation of target time characteristics.

STEP 3 - TRACK

A-15. During the track step, the target is observed and its activity and movement are monitored. Once the target is located and identified, maintain contact until an engagement decision is made and executed.

A-16. The track step begins once a definite fix is obtained on the target and ends when the engagement's results in the desired effect upon the target. Some targets may require continuous tracking upon initial detection as an emerging target. Sensors may be coordinated to maintain situational awareness or track continuity. Target windows of vulnerability should be updated when warranted. Relative priorities for information requirements are based on JFC guidance and objectives. TST generally receive the highest priority. If track continuity is lost, the fix step will likely have to be recompleted (and potentially the find step as well).

INPUTS

A-17. Inputs to the track step—
- Confirmed target.
- Target location and plot of movement (if applicable).

OUTPUTS

A-18. Outputs of the track step—
- Track continuity maintained on a target by appropriate sensor or combination of sensors.
- Sensor prioritization scheme.
- Updates to target window of vulnerability.

STEP 4 – TARGET

A-19. The target step takes an identified, classified, located, and prioritized target; finalizes the desired effect and targeting solution against it to include obtains required approval to engage. The target step can be time consuming due to the large number of requirements that must be satisfied. Timely decisions are more likely if target step actions can be initiated or completed in parallel with other steps.

A-20. The target step begins with target validation. Operations personnel ensure that an attack on the target complies with guidance, the law of war, and the rules of engagement. The target step matches available attack and sensor assets against the desired effect. Restrictions are resolved and the actions against the

Appendix A

target are coordinated and deconflicted. Risk assessment is performed before weapon systems selection. Weapon system is selected for engaging the target and assessment requirements are submitted.

INPUTS

A-21. Inputs to the target step—
- Identified, classified, located, and prioritized target.
- Restrictions for consideration are collateral damage estimation guidance, weapons of mass destruction, consequences of execution, law of war, rules of engagement, no-strike list, and restricted target list, component boundaries, and fire support coordination measures (FSCM).
- Situational awareness on available assets from all components.

OUTPUTS

A-22. Outputs of the target step—
- The desired effect is validated.
- Target data finalized in a format useable by the system that will engage it.
- Asset deconfliction and target area clearance considerations are resolved.
- Target execution approval (decision) in accordance with JFC and Service components commander's guidance.
- Assessment collection requirements are submitted.
- Consequence of execution prediction and assessment for weapons of mass destruction targets is performed.

STEP 5 - ENGAGE

A-23. During the engage step, the targets are confirmed as hostile and action is taken against the target.

A-24. The engagement is ordered and transmitted to the system selected to engage it. Engagement orders must be transmitted to, received by, and understood by those engaging the target. The engagement is managed and monitored by the engaging component and the desired result is successful action against the target.

INPUTS

A-25. Inputs to the engage step—
- Target approval decision.
- Selected engagement option.

OUTPUTS

A-26. Outputs of the engage step—
- Issuing and passing of the engagement order.
- Target engagement via scalable fires.
- Engagement direction and control.

STEP 6 – ASSESS

A-27. The assess step of dynamic targeting is the same as phase 6, the assessment phase of the joint targeting cycle. Both examine the results of the target engagement and the results of both must be integrated to provide the overall joint targeting assessment.

A-28. During the assess step, information is collected about the results of the engagement to determine whether the desired effects were achieved. TST or other HPT may require an immediate assessment to provide quick results and to allow for expeditious reattack recommendations.

INPUTS

A-29. Inputs to the assess step—
- Assessment requests matched against desired scalable fires.

OUTPUTS

A-30. Outputs of the assess step—
- Estimated or confirmed engagement results to decisionmakers in a timely manner.
- Reattack recommendations.

This page intentionally left blank.

Appendix B
Find, Fix, Finish, Exploit, Analyze, and Disseminate

Find, fix, finish, exploit, analyze, and disseminate (F3EAD) provides maneuver leaders at all levels with a methodology that enables them to organize resources and array forces across the range of Full Spectrum Operations. While the targeting aspect of F3EAD is consistent with the decide, detect, deliver, and assess (D3A) methodology, F3EAD provides the maneuver commander an additional tool to address certain targeting challenges, particularly those found in a counterinsurgency environment. F3EAD is not a replacement for D3A nor is it exclusive to targeting; rather it is an example of tactics, techniques, and procedures that works best at the battalion/tactical level for leaders to understand their operational environment and visualize the effects they want to achieve.

In counterinsurgency operations, targets assigned to nonlethal assets are frequently more important than targets assigned to lethal assets, and F3EAD is equally applicable for both. Effective targeting identifies options to support the commander's intent and objectives. For nonlethal effects, those options may include civil-military operations; information operations; and political, economic, and social programs. Lethal operations against targets are most typically designed to kill.

F3EAD is especially well suited and is the primary means for engaging personalities or high-value individuals (HVI). **A high-value individual is a person of interest (friendly, adversary, or enemy) who must be identified, surveilled, tracked and influenced through the use of information or fires. An HVI may become a high-payoff target (HPT) that must be acquired and successfully attacked (exploited, captured, or killed) for the success of the friendly commander's mission.** In this role, F3EAD features massed, persistent reconnaissance, or surveillance cued to a powerful and decentralized all-source intelligence apparatus to find a HVI in the midst of civilian clutter and find his exact location. This precise location enables surgical finishing operations (lethal or nonlethal) that emphasize speed to catch a mobile target. The emphasis on speed is not only to remove a combatant from the battlefield, but also to take the opportunity to gain more information on the foe. The exploit and analyze steps are often the main effort of F3EAD because these steps provide insight into the enemy's network and may open new lines of operation. The information accumulated during the exploit and analyze steps frequently starts the cycle over again by providing leads, or start points into the network that can be observed and tracked.

THE PROCESS WITHIN THE PROCESS

B-1. To gain an understanding of the F3EAD process, it is instructive to see how F3EAD is used within D3A and can begin during any phase of D3A methodology. The process still begins with a decide function in which decisions are made on priorities and the allocation of resources. The decide step is performed continuously, and requires extensive, persistent analytical work by operations and intelligence personnel. They analyze large quantities of all-source intelligence reporting to determine the following—
- Threat validity.
- Actual importance of potential targets.

Appendix B

- Best means to engage the target.
- Expected effects of engaging the targets (which will guide actions to mitigate negative effects).
- Any changes required to the exploitation plan.

B-2. As figure B-1 indicates, the detect function is broken into two parts, Find and Fix. During the find step, the HVI is identified and the target's network is mapped and analyzed. During the Fix step a specific location and time to engage the HVI is identified and the validity of the target is confirmed.

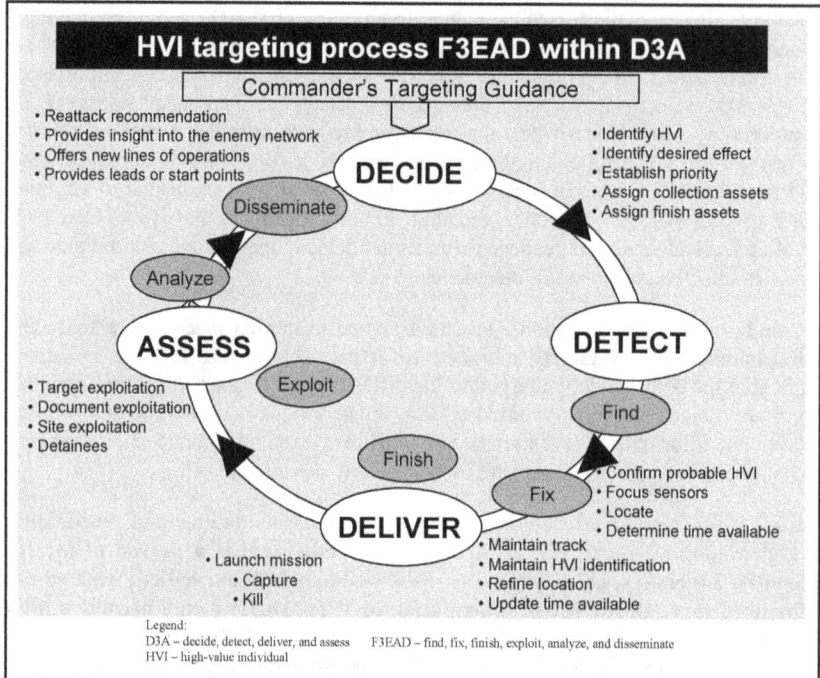

Figure B-1. High-value individual targeting process

B-3. In counterinsurgency operations, the target sets often will include HVI that require special care or caution in treatment because failure to attack them or to attack them improperly can lead to adverse consequences. As a result, the find step may take a considerable amount of time from days to even weeks as targets are identified and the target's network is mapped and analyzed.

B-4. The Finish step of F3EAD mirrors the deliver function of D3A when the action planned against the target is initiated and completed. Where the finish step differs from the deliver function in D3A is the nature of the means the commander will apply against identified target sets. In addition to systems typically associated with delivering effects in the D3A methodology, means used by a maneuver commander in counterinsurgency operations can include actions such as sniper, counter sniper, tactical callout, vehicle interdiction, and small ambush teams.

B-5. The exploit, analyze, and disseminate steps amplify the assess function. The engaging unit takes the opportunity to gather additional information during the exploit step, determines the implications and relevance of the information during the analyze step, and publishes the results during the disseminate step.

B-6. It is important to remember that the targeting process is continuous process. For any given target, the process tends to follow the flow depicted in figure B-1. At any given time however, a unit may be at the find step for some targets, the exploit step for several other targets, and at the fix, finish, analyze, or disseminate step for still other targets. Similarly, the unit may disseminate information pertaining to the location of a target prior to the finish step or exploit step. Generally, the process will follow the depicted flow, but don't let the process restrict what needs to happen next.

STEP 1: FIND

Effective operations are shaped by timely, specific, and reliable intelligence, gathered, and analyzed at the lowest possible level.

Field Manual (FM) 3-24

B-7. Collection plans need to be tailored to support the F3EAD process, driven largely by the requirement for time compression. Intelligence preparation of the battlefield (IPB) and area situational awareness collection will be an ongoing effort as day-to-day operations are conducted, but once an insurgent cell or HVI is identified, time is at a premium for building the target information folder. Once the target information folder is approved for collection, very specific information on possible take down locations must be collected quickly yet without alerting the HVI. This requires layering of collection efforts and assets.

B-8. HVI targeting will most often be conducted in counterinsurgency operations where the enemy frequently hides among civilian clutter. Persistent and high fidelity intelligence is often the key to defeating a foe whose primary strength is denying friendly forces a target. In contrast to major theater operations where the purpose is to find and destroy ships, tank formations, or infrastructure, the most difficult task in insurgencies is finding the enemy.

B-9. ISR assets are most effective against such enemies when massed. The insurgent's ability to hide in plain sight demands persistent collection in order to detect his presence. Persistent collection requires long dwell times and must be focused using multiple sensors on discrete parts of the network in order to achieve the fidelity of information required for targeting.

B-10. The effectiveness of reconnaissance and surveillance grows exponentially when it is cued to and driven by other sources of intelligence rather than operating alone. The enemy is so well hidden that it takes multiple sources of intelligence to corroborate information. Signals intelligence for example, can locate a target but may not be able to discern who it is. An airborne sensor with full motion video can track but not necessarily identify the target. Human intelligence can provide intent but may not be able to fix a target to a precise location. However, these disciplines working together are able to focus the spotlight on foes that are hidden in the general population, so they can be captured or killed. Without a robust, collaborative intelligence network to guide it, sensors are often used in reactive modes that negate their true power and tend to minimize their full potential. These intelligence disciplines provide a start point into the enemy network that can be exploited through persistent and patient observation. With this type of start point, one can mass reconnaissance forces with confidence that assets are not being wasted.

B-11. Inherent in massing is rejecting the commonly held practice of "fair sharing" intelligence among multiple units. Massing implies focus and priority. Selected parts of the enemy's network receive focus, which should be unwavering for a specified time. This is counterintuitive to those who feel the need to fair share assets as a way to cover more space and service more priorities. The problem with a low-contrast and fleeting foe, however, is that enemy actions are not easily predictable. Without prediction, the next best things are redundancy and saturation. The inability to mass employment of ISR assets over a large geographic area often results in a loss of effectiveness.

B-12. One tactic that is critical to improve effectiveness against an insurgent is nodal analysis (or link analysis). Insurgent networks do not exist in a vacuum. They interact with supporters in the population and, less directly, with their supporters buried in the power structure. HVI interacts with key leaders in politics, security, the economy, and real estate, as well as the general population.

B-13. Life pattern analysis is connecting the relationships between places and people by tracking their patterns of life. While the enemy moves from point to point, reconnaissance or surveillance tracks and notes every location and person visited. Connections between those sites and persons to the target are built, and nodes in the enemy's network emerge. Link analysis and life pattern analysis identify these relationships in order to flesh out the target information folder. To be effective, there must be sufficient intelligence on the network the HVI belongs to in order to know the effect of his removal. Just because he

Appendix B

is the cell leader may not be a good enough reason to target him. How will the cell be hurt by his removal? How long will it take to replace him?

B-14. This analysis has the effect of taking a shadowy foe and revealing his physical infrastructure for things such as funding, meetings, headquarters, media outlets, and weapons supply points. As a result, the network becomes more visible and vulnerable, thus negating the enemy's asymmetric advantage of denying a target. Nodal analysis uses the initial start point to generate additional start points that develop even more lines of operation into the enemy's network. The payoff of this analysis is huge but requires patience to allow the network's picture to develop over a long term and accept the accompanying risk of potentially losing the prey.

B-15. Networks are notably resistant to the loss of any one or even several nodes. The focus of targeting is not just to identify an individual who is a leader in the network. Instead, it is to identify the critical leader whose removal will cause the most damage to the network. The ultimate success is to remove sufficient critical nodes simultaneously—or nearly so—such that the network cannot automatically reroute linkages, but suffers catastrophic failure.

INPUTS

B-16. Inputs to the find step—
- Commander's guidance and priorities.
- IPB, to include identified named areas of interest, target areas of interest, and cross cueing of intelligence disciplines to identify potential target sites or operational environments.
- Life pattern analysis.
- Collection plans based on the IPB.

OUTPUTS

B-17. Outputs of the find step—
- Potential HVI detected and nominated for further development.
- Target folders.
- HVI network identified and analyzed.

STEP 2: FIX

B-18. The continued collection effort paints a picture of the HVI. The intelligence staff officer (S-2) can draw broad behavior patterns that will focus the specific collection requirements from analysis of the intelligence. The information harvested from the focused and persistent collection reveals the life patterns of the HVI includes overnight locations, daily routes, visitations, and trustworthy associates. National and unit intelligence assets then corroborate the life patterns. As the details are filled in, it becomes possible to anticipate where the HVI is most likely to spend time or visit.

B-19. Not only does this intelligence provide the unit a more complete understanding of the network, it also helps the unit to confirm that the planned action will have the desired result and be worth the cost. Sometimes, intelligence gained from continuing to monitor the HVI is more significant than killing or capturing the HVI. An action against one target may reduce the chance of success against a more important target.

B-20. Maintaining persistent, continuous intelligence support is particularly hard at lower echelons of command and small units where intelligence assets are less available than at the brigade combat team (BCT) and higher commands. In these units, it is important for the command to establish intelligence support teams with personnel who know the targets and are trained in the unit standard/standing operating procedures (SOP) for sensor preparation/briefings, patrol debriefings, data collection, and able to fuse this information with the unit's operational plan to finish the target.

B-21. As the probable location of the HVI target is narrowed to a few sites, the unit is able to identify feasible courses of action and begin refining the planned actions of the finish force. At some point the information leads the unit to determine a HVI target is likely to be a specific location (fix) at a specific time or within a specific time frame. Depending on the accuracy and reliability of the information, the unit may chose to verify the information through other means. Once the unit is satisfied that the fix is valid, they may chose to launch the finish force.

INPUTS

B-22. Inputs to the fix step—
- Probable HVI.
- Information on the target and the target's network.

OUTPUTS

B-23. Outputs of the fix step—
- Target identification and confirmation.
- Target location accuracy refined to the level required for target engagement.
- Determination or estimation of target time characteristics.

STEP 3: FINISH

B-24. The window of opportunity to engage the target requires a well trained and rehearsed finish force and a well developed SOP. The force will normally not have the time to create elaborate plans. Instead, the force will be required to adapt a known drill to the existing conditions and rapidly execute the required actions, such as a raid, ambush, or cordon and search. The force must also be prepared to conduct follow-on operations based on information found during exploitation on the objective.

INPUTS

B-25. Inputs to the finish step—
- HVI location within a given time frame.

OUTPUTS

B-26. Outputs of the finish step—
- Target isolated and engaged.
- Target location secured.
- Exploitation force on site.

STEP 4: EXPLOIT

B-27. Once secured, the target site must be exploited. Site exploitation is a methodical, detailed collection process to gather potential intelligence. Effective site exploitation requires prior planning to include SOP, search plans, prepared site exploitation kits, and tactical questioning plans. Units must make these preparations in advance of the finish step in order to be conduct effective actions on the objective.

B-28. The site exploitation team may have a variety of enablers in direct support, or it may come solely from the unit. In any case, they must have clear instructions on what to look for in the specific site and training in how to conduct the investigation. Some units use "smart cards" with target specific information and predetermined questions. Such aids have been useful in preparing and guiding the exploitation teams. Some organizations prefer designated assault/exploitation units. Continual preparation for these type missions allows the development and refinement of SOP.

B-29. F3EAD differs from other targeting models because of its emphasis on the exploit and analyze steps as the main effort. This recognizes the importance of intelligence in fighting the low contrast foe and aggressively supplying multisource start points for new information collection. More than the other steps, this feeds the intelligence operations cycle in which intelligence leads to operations that yield more intelligence leading to more operations. The emphasis on raids is essential to gather intelligence on the enemy network; simply killing the enemy will not lead to greater effectiveness against their networks. In fact, capturing the enemy for purposes of interrogating is normally the preferred option.

B-30. Target exploitation and document exploitation are important law enforcement activities critical to F3EAD. Documents and pocket litter, as well as information found on computers and cell phones, can provide clues that analysts need to evaluate enemy organizations, capabilities, and intentions. The enemy's network becomes known a little more clearly by reading his email, financial records, media, and servers. Target and document exploitation help build the picture of the enemy as a system of systems.

B-31. The interrogation of detainees is crucial to revealing the enemy's network. The ability to talk to insurgent leaders, facilitators, and financiers about how the organization functions offers significant insight on how to take that organization apart. Intelligence from detainees drives operations, yielding more detainees for additional exploitation and intelligence.

INPUTS

B-32. Inputs to the exploit step—
- Secured target location.
- Targeted questions.
- Site exploitation preparation and SOP.

OUTPUTS

B-33. Outputs of the exploit step—
- Documented information.
- Detailed reports.
- Follow on targets for immediate execution.

STEP 5: ANALYZE

B-34. The bottom line of the analyze step is to examine and evaluate information and rapidly turn it into actionable intelligence that can be applied to defeat the enemy's network. Some information may be immediately actionable, such as information providing the location of another HVI. Other information may need further analysis and corroboration.

B-35. The avalanche of information requires the staff to streamline operations to allow for this data to be stored, analyzed, recalled, and disseminated as necessary. New or additional players must be included in the collection and assessment process. National and theater level technical assets will also be critical and mechanisms to facilitate their integration must be developed. All of this will require modifications of existing planning mechanisms and procedures, and learning how to incorporate new sources.

B-36. The objective is to make intelligence, not information. To do this you have to invest resources and focus on preparation. The level of dedicated resources (mainly personnel) will have a direct correlation to the quality and quantity of developed intelligence. Too few resources result in an extrication of raw information effort, instead of an analytical and understanding effort. The more investment the greater the return.

INPUTS

B-37. Inputs to the analyze step—
- Document information.
- Detailed reports.

OUTPUTS

B-38. Outputs of the analyze step—
- Actionable intelligence.

- Correlated information.
- Intelligence assessments.

STEP 6: DISSEMINATE

B-39. The sisseminate step is simple but time consuming. The goal is to make sure everyone else knows what you know. Even information that appears to be irrelevant may hold the key to unlocking a network for someone else. Fortunately, the various computer programs and networks greatly aid the dissemination process.

B-40. Prioritizing the dissemination effort is essential. Some information will answer a priority intelligence requirement (PIR) and should be forwarded to the requesting agency immediately. Other information may be important based on the operational environment. Still other information will be routine and can be handled routinely.

INPUTS

B-41. Inputs to the disseminate step—
- Relevant and correlated information.
- Actionable intelligence.
- Intelligence assessments.

OUTPUTS

B-42. Outputs of the disseminate step—
- Databases, matrices, and assessments are updated.
- Intelligence and information is pushed to higher, lower, and adjacent units.
- Information is made available to everyone with a need to know.

MEASURING SUCCESS

B-43. Measuring success when conducting F3EAD targeting requires analysis conducted in two stages. The first stage occurs immediately after the finish step and should answer questions associated directly to the target and its network. Examples of first stage metrics include—
- Killed or captures insurgents.
- Changes in insurgent patterns.
- Captured equipment and documents.

B-44. The second stage of analysis takes the longer view. These metrics provide the yardstick for JFC to examine progress made toward meeting objectives established in the joint campaign plan to include—
- Changes in local attitudes towards United States and host nation forces to include public perceptions.
- Changes in the quality or quantity of information provided by individuals or groups.
- Changes in the economic or political situation of an area.

This page intentionally left blank.

Appendix C
National Intelligence Organizations Support to Targeting

Many intelligence organizations within the United States Government are tasked with collecting information on potential threats against friendly interest. Only a portion of the information is relevant to any given operation. Successful military operations require information that is fused and focused.

Combatant command joint intelligence operations center possess organizational processes to integrate and synchronize military, national, operational, and tactical intelligence capabilities to increase intelligence fidelity and timeliness of dissemination to Service components, and to decrease duplication of effort by intelligence centers.

The intelligence directorate of a joint staff (J-2) assesses the combatant command's organic tasking, collection, processing, exploitation, and dissemination capabilities to support the combatant command's selected operations through all phases of conflict. The combatant command J-2 determines intelligence shortfalls and, working with the Defense Joint Intelligence Operations Center (DJIOC), begins to establish federated partnerships with other intelligence organizations to address these shortfalls. Federated partnerships are formal agreements with other theater joint intelligence operations centers, Service intelligence centers, defense intelligence agencies, reserve intelligence elements, or other nongovernmental intelligence agencies to assist with the combatant command J-2 intelligence responsibilities.

In a "federated approach," a joint force commander (JFC) receives its principal intelligence support from the combatant command's joint intelligence operations center, which receives information from all echelons and performs all-source analysis and production. The following paragraphs contain information on the organizations that may provide expertise for federated intelligence support to targeting and allow access to more actionable information than would otherwise be available to JFC.

DEPUTY DIRECTOR FOR TARGETING, JOINT STAFF INTELLIGENCE DIRECTORATE

C-1. The Deputy Directorate for Targeting, Joint Staff Intelligence Directorate is a unique organization since it is a major component of the Defense Intelligence Agency, which is a combat support agency, as well as a fully integrated element of the joint staff. The J-2 is the primary coordination element for national level intelligence support to joint targeting.

C-2. The deputy directorate for targeting functions as the lead agent for providing and coordinating national level intelligence support to joint targeting. Specific deputy directorate for targeting responsibilities include—
- Providing the Chairman of the Joint Chiefs of Staff and the operations directorate of a joint staff (J-3) with targeting, assessment, and technical support during contingency and crisis action planning.
- Providing the combatant commands, if requested and validated, with intelligence community target development through all phases of the targeting cycle.

Appendix C

- Assisting the combatant commands in establishing, coordinating, or supporting federated intelligence operations, to include target development and assessment.
- Assisting combatant commands with coordination of intelligence community target vetting.
- Providing functional expertise on targeting and targeting related issues undergoing Joint Staff, Secretary of Defense, and Presidential review. This includes, but is not limited to, command target lists, planning orders, warning orders, and sensitive target and review products.

Note. See Joint Publication (JP) 2-0 for additional details.

DEFENSE INTELLIGENCE AGENCY, DEFENSE JOINT INTELLIGENCE OPERATIONS CENTER

C-3. The Director of the Defense Intelligence Agency serves as the Chief, DJIOC and reports to the Secretary of Defense through the Chairman of the Joint Chiefs of Staff. The Defense Intelligence Agency provides finished target intelligence to the President, Secretary of Defense, and JFC, providing worldwide support across the entire spectrum of conflict.

C-4. The DJIOC plans, prepares, integrates, directs, synchronizes, manages continuous full spectrum Department of Defense intelligence operations in support of the combatant commands, and is the primary conduit through which national level target intelligence support is provided to the combatant commands and subordinate joint forces. This includes targeting, information operations, battle damage assessment (BDA), current intelligence, and target systems analysis of the adversary. The DJIOC coordinates and prioritizes military intelligence requirements across the combatant commands, combat support agencies, Reserve Components, and Service component intelligence centers.

C-5. DJIOC is responsible for providing target intelligence to the President of the United States or Secretary of Defense, combatant commanders, and joint task force commanders in support of joint worldwide operations. The DJIOC directly supports Joint Staff J-2 targeting efforts by consolidating all-source target development and material production. The DJIOC and combatant command JIOC controls national intelligence assets and determine requirements through the Director of National Intelligence and intelligence community representatives to combatant commands.

Note. For additional details, see JP 2-01.

JOINT FUNCTIONAL COMPONENT COMMAND FOR INTELLIGENCE, SURVEILLANCE, AND RECONNAISSANCE

C-6. The Joint Functional Component Command for Intelligence, Surveillance, and Reconnaissance (JFCC-ISR) plans, integrates, and coordinates defense global ISR strategies in support of joint operation planning and combatant command planning/operations in accordance with the United States Strategic Command's combatant command plan assigned ISR mission. The Joint Functional Component Command for Intelligence, Surveillance, and Reconnaissance formulates recommendations to integrate global ISR capabilities associated with the missions and requirements of Department of Defense ISR assets in coordination with the DJIOC and Commander, United States Strategic Command. The Joint Functional Component Command for Intelligence, Surveillance, and Reconnaissance provides personnel and resources in direct support of the combatant command joint intelligence operations center.

NATIONAL SECURITY AGENCY

C-7. The National Security Agency provides critical intelligence support to all phases of joint targeting. The National Security Agency's Information Warfare Support Center serves as the agency's primary point of contact for organizations seeking specific targeting or targeting related analytical information. The Information Warfare Support Center directly assists with the preparation of information operations strategies as well as all-source targeting studies for the Department of Defense, Joint Staff, combatant commands, and joint task forces. This support includes analysis of communications networks or other aspects of the information infrastructure, as well as operational Signals Intelligence. The National Security

Agency is also responsible for providing the combatant command, Joint Staff J-2, and Defense Joint Intelligence Operations Center with the intelligence gain or loss assessment, which is an evaluation of the quantity and quality of intelligence data lost if desired effects are created on a target. The National Security Agency is also tasked with keeping the Defense Joint Intelligence Operations Center, combatant command Joint Intelligence Operations Centers, and other interested command and agencies informed of agency activities that take place in each respective combatant commander's area of responsibility or subordinate joint forces' operational area.

NATIONAL GEOSPATIAL-INTELLIGENCE AGENCY

C-8. The National Geospatial-Intelligence Agency is a Department of Defense combat support agency, as well as a national intelligence organization. The agency provides tailored mapping products, services, and training support to the Department of Defense, Joint Staff, combatant commands, and joint task forces. Mapping products use geodetically controlled source material and refined mensuration techniques and data. Major targeting assistance includes the digital point positioning database and the mensuration of precise points to support targeting. The National Geospatial-Intelligence Agency is the central authority responsible for managing imagery intelligence and is the primary source for geospatial intelligence analysis and products at the national level. For this reason, their staff plays a critical role providing collection support to target intelligence efforts. The DJIOC validates all national imagery nomination requests, deconflicts multiple requirements, and implements tasking of national imagery assets.

C-9. The National Geospatial-Intelligence Agency will, when requested, provide geospatial intelligence support to the combatant command via a support team or as part of a national intelligence support team. Their support teams are established at each combatant command headquarters and are in direct support to the combatant command joint intelligence operations center. The support team provides the full spectrum of National Geospatial-Intelligence Agency's geospatial intelligence capabilities and is composed of a core cadre that includes geospatial analysts, imagery analysts, and staff officers. The support team also has full connectivity with the agency to ensure reachback capability into their total support effort. A National Geospatial-Intelligence Agency support team would contribute to responsive imagery tasking, collection, processing, exploitation, and dissemination in support of joint targeting efforts.

> *Note.* For more on geospatial intelligence capabilities, target support products, and services see JP 2-03 and JP 3-60.

JOINT INFORMATION OPERATIONS WARFARE COMMAND

C-10. The Joint Information Operations Warfare Command plans, integrates, and synchronizes information operations in support of JFC and serves as the United States Strategic Command lead for enhancing information operations across the Department of Defense. It exists to provide the full range of information operations options to the supported commander, focusing on the operational level of war, but prepared to support tactical and strategic level requirements as well.

C-11. The Joint Information Operations Warfare Command supports the JFC by conducting following tasks—

- Supports the integration of operational security, military information support operations, military deception, public affairs, electronic warfare (EW), and destruction throughout the planning and execution phases of an operation.
- Interfaces with the Joint Staff, Services, Department of Defense, and intergovernmental agencies to coordinate and integrate information operations efforts for joint task force commanders.
- Participates in joint special technical operations support to combatant commanders.
- Provides planning consideration guides and precision influence target folders to supported combatant commands.
- Coordinates and integrates the information operations portion of the intelligence preparation of the operational environment.
- Evaluates information operations effectiveness in military operations.

Appendix C

- Provides communications system and intelligence nodal analysis and information operations targeting support.
- Assists with strategic information operations planning and engagement.

JOINT WARFARE ANALYSIS CENTER

C-12. The Joint Warfare Analysis Center assists in preparation and analysis of joint operation plan (OPLAN) and Service chiefs' analysis of weapons effectiveness. It provides the Joint Staff, combatant commands, joint force commanders, and other Department of Defense and intergovernmental agencies with precision targeting and deterrent options for selected networks and nodes. The Joint Warfare Analysis Center provides specialized analysis for use in developing targeting strategies. The analysis includes innovative and accurate engineering and modeling based targeting options and helps provide planners with an understanding of risks and consequences, including collateral damage estimates. They normally provide this support to the joint task force through the supported combatant command.

DEFENSE THREAT REDUCTION AGENCY

C-13. The Defense Threat Reduction Agency is a combat support agency charged with developing methods to deal more effectively with threats by chemical, biological, radiological, nuclear, and high-yield explosives weapons of mass destruction and preventing future threats. It covers a broad spectrum of activities, but is directly involved in the targeting process by making collateral damage and casualty estimations when employing weapons against facilities that may contain weapons of mass destruction. The Defense Threat Reduction Agency provides target characterization and high fidelity weapons effects modeling to support physical and functional defeat of hardened and deeply buried targets.

C-14. The Defense Threat Reduction Agency—
- Maintains continuous global situational awareness of weapons of mass destruction to support decisive action.
- Provides hazard predictions and consequence management expertise.
- Develops technologies and tactics, techniques, and procedures to hold at risk and defeat critical military targets protected in tunnels and other deeply buried, hardened facilities.
- Provides the Department of Defense nuclear mission support.
- Provides enhanced capabilities to assess enemy weapons of mass destruction operations.

JOINT SPACE OPERATIONS CENTER

C-15. The Joint Space Operations Center is the primary United States Strategic Command interface for coordinating and delivering joint space effects to the supported commander, to include all aspects of joint operation planning and the air tasking cycle. The Joint Space Operations Center is responsible for analyzing and targeting enemy space capabilities in support of theaters. Joint Space Operations Center targeteers can evaluate theater Air Operation Directives and nominate specific space related targets to meet a theater commander's objectives.

C-16. The primary functions of the Joint Space Operations Center are to—
- Develop a global space operations strategy to meet Commander, United States Strategic Command objectives and guidance.
- Assist development of theater space operations strategy to meet the geographic combatant commander objectives and guidance through robust interaction with theater space coordination authority.
- Produce and disseminate the joint space tasking order.
- Task and execute day-to-day space operations for assigned and attached space forces.
- Receive, assemble, analyze, filter, and disseminate space related all-source intelligence and weather information to support air and space operations planning, execution, and assessment.

- Conduct operational level assessments to determine mission and overall space operations effectiveness as required by the Commander, United States Strategic Command, and other geographic combatant command to support global and theater combat assessments.

Note. See JP 3-14 for additional details.

JOINT TECHNICAL COORDINATING GROUP FOR MUNITIONS EFFECTIVENESS

C-17. Joint Technical Coordinating Group for Munitions Effectiveness is a vital joint service activity that develops operational effectiveness estimates for all nonnuclear munitions and munitions effective miss distance tables that contain collateral damage distances for all air-to-surface and surface-to-surface conventional munitions, and continuously updates Joint Munitions Effectiveness Manuals used by the Services for training and tactics development, operational targeting, weapons selection, aircraft load outs, and planning for ammunition procurement, survivability, and development of improved munitions. The Joint Technical Coordinating Group for Munitions Effectiveness directs the analytical effort of working groups necessary to determine degrading effects of various terrain environments on nonnuclear munitions effectiveness and improving the database for target vulnerability, delivery accuracy, and weapons characteristics.

NATIONAL AIR AND SPACE INTELLIGENCE CENTER

C-18. The National Air and Space Intelligence Center is the sole national center for integrated intelligence analysis on air and space systems, forces, and threats. It assesses current and projected foreign air and space capabilities and intentions, develops targeting and mission planning intelligence materials, and evaluates evolving technologies of potential adversaries. Such technical information is useful in determining how to create specific effects on specific targets and target systems. In addition to expertise on worldwide air assets, the National Air and Space Intelligence Center also has leading experts on long-range surface-to-surface missiles (such as medium range and intercontinental ballistic missiles).

C-19. The National Air and Space Intelligence Center can provide target systems analysis of—
- Communications system and intelligence.
- Air forces and airfields.
- Integrated air defense forces.
- Space forces.
- Ballistic missile forces.

UNITED STATES JOINT FORCES COMMAND QUICK REACTION TEAM

C-20. The United States Joint Forces Command quick reaction team is a rapidly deployable team of targeting personnel and collection managers designed to provide immediate crisis support to combatant commands. The quick reaction team can deploy within 24 hours at the request of a combatant commander via Joint Staff J-2 or the Defense Joint Intelligence Operations Center. They are trained analysts, but must be integrated into existing theater intelligence organizations as they deploy with no organic automated data processing or communications support. The supported combatant commander determines the team's location within theater (headquarters, joint intelligence operations center, joint task force, or component command) based on assessed needs. The quick reaction team is not a permanent targeting or collection augmentation and should be returned to national control as mobilization or individual augmentation arrive to support the combatant commander's requirements.

CENTRAL INTELLIGENCE AGENCY

C-21. The Central Intelligence Agency, through its target support group within its Office of Military Affairs works closely with the Department of Defense on many issues relating to every phase of the targeting cycle. The target support group makes a variety of Central Intelligence Agency resources available to military target planners. They can provide target systems analysis of communications system and intelligence, weapons of mass destruction, and counterterrorism. Additionally, in peacetime, applicable request for information are routed to the Central Intelligence Agency to be addressed by the agency's Office of Military Affairs. The target support group provides information and expertise in support of military target development and processes formal requests for target nominations (review and approval by the agency's leadership) to add the Central Intelligence Agency selected targets to a Department of Defense plan. The target support group manages all military Special Technical Operations and Special Access Program compartments, and deconflicts military targeting with Central Intelligence Agency operational assets. In a crisis or war, Central Intelligence Agency personnel or teams can be attached to combatant commands, joint task forces, or joint force components, as required.

DEPARTMENT OF STATE, BUREAU OF INTELLIGENCE AND RESEARCH

C-22. The central point of contact within the Department of State for intelligence, analysis, and research is the Bureau of Intelligence and Research. The Bureau of Intelligence and Research produces intelligence studies and analyses, which have provided valuable information in support to targeting. As the lead foreign affairs agency and the enabler of United States diplomacy, the State Department has a unique perspective on the nations of the world. All-source reporting via Foreign Service channels at American embassies or consular posts has also proven useful, particularly during the objectives and guidance, target development, and combat assessment phases of the targeting cycle. Intelligence concerning political and military leaders, cultural trends and thoughts, and economics—to name just a few areas—can provide information that ties military strategy to the entire spectrum of national power. Even from a purely military standpoint, such intelligence can enhance understanding of adversary motivations.

Appendix D
Example Formats and Target Reports

The targeting products developed during the targeting process are actually tools. The commander, the targeting working group, supporting, and supported units use them. The products allow them to control and synchronize targeting in an effective and efficient way. There are no prescribed formats. Each unit will develop tools that work best for them. Factors to consider in developing formats are as follows—

- Type and level of the command.
- Operating environment.
- Assets available.
- Missions.
- Standard/standing operating procedures (SOP).

Regardless of the formats used, the decide, detect, deliver, and assess (D3A) methodology associated with the command decision cycle must be followed. The quality of targeting products is paramount to the commander's and staff's ability to attain and maintain credibility with other warfighting functions. Targeting products can be presented in many forms. These forms may be oral presentations, hard copy publications, or electronic format.

The purpose of this appendix is to provide a menu of formats and a focus on the targeting information and knowledge the commander and staff requires. The formats may be copied or modified by the targeting working group to support requirements of the command.

HIGH-PAYOFF TARGET LIST

D-1. The modified high-payoff target list (HPTL) (see figure D-1 below) is a sample of the basic format described in Chapter 2.

Appendix D

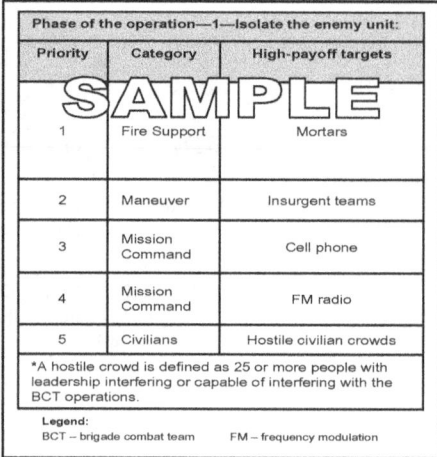

Figure D-1. High-payoff target list (sample)

TARGET SELECTION STANDARDS

D-2. Target selection standards (TSS) (see the sample format in figure D-2 below) are usually comprised of the essential elements listed below units may develop their own target selection format—

- High-payoff target (HPT). This refers to the designated HPT that the collection manager is tasked to acquire.
- TIMELINESS. Valid targets are reported to weapon systems within the designated timeliness criteria.
- ACCURACY. Valid targets must be reported to the weapon system meeting the required target location error (TLE) criteria. The criteria are the least restrictive TLE considering the capabilities of available weapons system.

High-payoff targets	Timeliness	Accuracy
COPs	3 hrs.	500 m
RISTA	30 min.	150 m
Rockets	30 min.	500 m
Missile Attacks	30 min.	500 m
Air Defense Artillery	15 min.	500 m
Command Post	3 hrs.	500 m
Ammunition	6 hrs.	1 km
Maneuver	1 hr.	150 m

Legend:
COP – communications operational planning system   RISTA – reconnaissance, intelligence, surveillance, and target acquisition

Figure D-2. Target selection standards matrix (sample)

Example Formats and Target Reports

ATTACK GUIDANCE MATRIX

D-3. The attack guidance matrix (AGM) (see sample figure D-3) provides guidance on what HPT should be attacked and when and how they should be attacked. Units may develop their own AGM format. The AGM consists of the following elements—

- HPTL. The HPTL is a prioritized list of HPT by phase of the operation.
- WHEN. The WHEN column indicates the time the target should be engaged. (See the legend below the example.)
- HOW. This column indicates the weapon system that will engage the target.
- EFFECT. The desired effects on the target or target system are stated in this column.
- REMARKS. Remarks concerning whether or not battle damage assessment (BDA) is required, whether coordination must take place, and so forth are indicated in this column.

High-payoff target	When	How	Effect	Remarks
Mortars	I	Field Artillery	Destroy	Use search and attack teams in restricted areas
Insurgent teams	I	Field Artillery	Neutralize	Destroy mission command
Cell phone	A	Electronic Attack	Disrupt	Disrupt service starting H-2
FM radio	A	Electronic Attack	Disrupt	No jamming until H-3 to preserve intelligence
Hostile civilian crowds	A	MISOP/MP	Dispersed	25 or more with leadership constitute crowd

Legend:
(A) - as acquired FM – frequency modulation H - hour (H-hour is the time for a scheduled event to begin.)
(I) – immediate MISOP – military information support operations MP – military police

Figure D-3. Attack guidance matrix (sample)

THE HIGH-PAYOFF TARGET LIST AND ATTACK GUIDANCE MATRIX

D-4. In the samples below, there are two different formats for HPTL and associated AGM. Units may develop their own HPTL and AGM formats. Table D-1 is a straightforward combined format compared to table D-2. This allows the targeting working group to specify HPT in priority order with as much detail as desired. It also allows the team to immediately specify the **when, how,** and **restrictions** information for attack of the HPT. The HPTL and AGM will likely change as the situation changes from one phase or critical event to another. Therefore, a separate HPTL and AGM can be prepared for each phase of the battle.

Appendix D

Table D-1. HPTL-AGM option 1 (sample)

Event or Phase: Attack through security zone	
High-payoff target list	Attack guidance – when, how and restrictions
COPS	Prep, A, N cannon, and/or rocket
RISTA	Prep, A, N cannon, and/or rocket
Rockets	Prep, I, N cannon, and/or rocket
Missile attacks	Prep, I, N cannon, and/or rocket Use corps assets beyond PL Diamond
Air Defense Artillery	SEAD. R, S
Command post	Prep, A, D
Ammunition sites	Prep, A, D
Maneuver and recon patrols	Prep, A, N
Legend: WHEN (I) = Immediate EFFECTS (S) = Suppress WHEN (A) = As Acquired EFFECTS (N) = Neutralize WHEN (P) = Planned EFFECTS (D) = Destroy COPS – communications operational planning system PL – phase line Prep – preparation fires RISTA – reconnaissance, intelligence, surveillance, target acquisition SEAD – suppression of enemy air defense	

Table D-2. HPTL-AGM option 2 (sample)

High-payoff target	When	How	Effect	Remarks
Insurgent Mortars	I	Field Artillery	Destroy	Use search and attack teams in restricted areas
Insurgent Teams	I	Field Artillery	Neutralize	Destroy mission command
Insurgent Cell Phone	A	Electronic Attack	EW	Disrupt service starting H-2
Insurgent FM Radio	A	Electronic Attack	EW	No jamming until H-3 to preserve intelligence
Hostile Civilian Crowd	A	MISOP/MP	Dispersed	25 or more with leadership constitute crowd

Legend:
(A) - as acquired   EW – jamming   FM – frequency modulation
H - hour (H-hour is the time for a scheduled event to begin.)   (I) - immediate
MISOP – military information support operations   MP – military police

COMBINED HIGH-PAYOFF TARGET LIST, TARGET SELECTION STANDARDS, ATTACK GUIDANCE MATRIX

D-5. The doctrinal version of the AGM can be modified to make it more useful and practical. The format below combines all three formats discussed in Chapter 2 into one format. The modified HPTL-TSS-AGM is focused on exclusively attacking HPT. It can be completed in minutes and can be formatted in the maneuver control system to allow for immediate transmission to all who need it. (See figure D-4.)

D-6. The CATEGORY column on the AGM is too generic to be useful. In the combined version, specific HPT (according to phase, echelon, and so forth) are listed across the top. This sends a clear message that only HPT will be attacked. If we accept the premise that the destruction of HPT will defeat the enemy (for example, preclude successful completion of his mission), then we must concentrate our limited resources only on HPT.

D-7. Down the left side, the weapon systems available to brigade, division, and corps are listed. This allows for quick reference to determine which assets are available.

D-8. The WHEN column on the old AGM is unnecessary when attacking only targets identified as HPT. The distinction between the terms as acquired and immediate becomes blurred to the point that they are indiscernible.

D-9. HOW column loses most of its utility because the terms suppress and neutralize are too subjective for commanders and fire support officers (FSO). As previously stated, if we are focused solely on HPT, the destruction of those HPT is what will result in the defeat of the enemy.

D-10. In the modified version of the matrix each block contains the TSS, numbered 1 through 4, for the following information:
- Required TLE is given meters.
- Required target size is described in type of units.
- Activity of the target is described by moving or stationary.
- Time of acquisition expresses the first actual spotting.

D-11. When these criteria are met, the applicable weapon system(s) is notified to engage the target. In the case of targets that qualify for attack by more than one weapon, the weapon systems are prioritized. The priority is listed in the upper right corner of the matrix block. (In the completed HPTL-TSS-AGM below, fires brigade is the second priority for attacking the field artillery units listed as HPT number 1.)

D-12. The REMARKS column allows for the discussion of restrictions, constraints, or restraints involving HPT or weapon systems. (For example, the Army tactical missile systems may only be used on semi-fixed or soft-fixed targets.)

D-13. The phase of the operation and an effective date time group is included to eliminate confusion over which version is current.

Appendix D

Phase: II					DTG: 030530 Jun02	
HIGH-PAYOFF TARGETS						
Priority		1	2	3	4	Remarks
Descriptions		Field Artillery	ADA	Maneuver	Command Post	
A T T A C K S Y S T E M	Fires Bn 1 100 m 2 Btry 3 Stat 4 1 hr	5	2	3	3	
	Fires Bde 1 100 m 2 Btry 3 Stat 4 1 hr	2	1	2	2	
	MLRS 1 100 m 2 Btry 3 Stat 4 1 hr	1	1	2	2	
	ATACMS 1 100 m 2 Btry 3 Stat 4 1 hr	1	1	1	1	Must request from corps
	MNVR 1 1 km 2 Btry 3 Stat/Move 4 1 hr	6	5	6	6	
	ATK Aircraft 1 500 m 2 Btry 3 Stat 4 1 hr	3	3	4	4	
	CAS 1 500 m 2 Btry 3 Stat 4 1 hr	4	4	5	5	

Reference: Target Selection Standard
1. Required TLE
2. Size
3. Activity
4. Time acquired
   Priority of attack

Legend:
ADA – air defense artillery ATK– attack ATACMS – Army tactical missile system
Bde – brigade Bn – battalion Btry – battery
CAS – close air support CP – command post MLRS– multiple launch rocket system
MNVR – maneuver Stat - stationary

Figure D-4. Combined HPTL TSS AGM (sample)

TARGET SELECTION STANDARDS WORKSHEET

D-14. The sample form below (figure D-5) incorporates TSS into a document that can be used to track and confirm or deny targets generated by each sensor source. Units may develop their own target selection standards worksheet format. The column headings are described below—
- HPT. This column is used to list HPT.
- SOURCE. This column is used to list the particular sensor agent.
- TARGET LOCATION. This column is used to record the target by grid location.
- ACCURACY (Target Location Error). This column lists the reliability of the sensor, normally stated in meters.

Example Formats and Target Reports

- TIME OF TARGET. This column is used to record the date time group the sensor acquired the target.
- TIME LIMIT. This column is used to tell the staff how old the acquisition can be and still be attacked.
- VALIDITY CONFIRMED. In this column, using YES or NO to record any confirmation by a second source. Confirmation by another sensor may not be necessary depending on the sensor.
- CLEARANCE CLEARED. This column is used to record who or what agency cleared the target for attack. This is especially critical where the potential for fratricide exists.

TARGET SELECTION STANDARDS WORKSHEET							
High-payoff target	Source	Target Location	Accuracy (Target Location Error)	Timeliness		Validity Confirmed	Clearance Cleared
				Time of Target	Time Limit		

Figure D-5. Target selection standards worksheet (sample)

TARGETING SYNCHRONIZATION MATRIX

D-15. The sample targeting synchronization matrix has been successfully used to synchronize targeting by assigning responsibilities to detect, deliver, and assess attacks on specific HPT. The HPT is listed in priority by category under the DECIDE column. Units and agencies are listed under the DETECT, DELIVER, and ASSESS columns across from the specific HPT for which they are responsible. As responsibilities are fixed, the asset envisioned to be used is also indicated. This provides the targeting working group the checks to ensure all assets are used and that assets or agencies are not overtaxed. This form could also be prepared for a specific event or for each phase of the battle. Units may develop their own targeting synchronization matrix format (See figure D-6).

Appendix D

P	Category	HPT	Decide	Detect		Deliver		Assess	
			Agency	Agency	Asset	Agency	Asset	Agency	Asset
1	Fire Support	Mortars	Fires Bde		Fire Finder	1-Fires Bde 2-Avn	1-Arty MLRS	Avn	INFLTREP
		Rocket Artillery	G2		EAD assets			G2	Analysis
		Tactical Ballistic Missiles	313 MI		Quick fix			313 MI	Quick fix
		Theater Missile	Fires		Fire Finder	Fires Bde	Arty, MLRS		
		Atk Aircraft	3-4 ADA		Organic National	3-4 ADA Fires Bde	Organic Arty, MLRS	3-4 ADA	Organic
2	ADA	Enemy Air Theater Missile	G2		Tactical Exploitation of National Capabilities	Fires Bde	SEAD Arty, MLRS	G2	Tactical Exploitation of National Capabilities
					EAD	G3	Data linked aircraft		EAD ELINT
3	Recon	Launch Point				Fires Bde	Arty, MLRS	Air Force	INFLTREP
					EAD ELINT			G2	EAD ELINT
		Patrols OPs			Organic	Bde	Organic	Bde	Organic

Legend:
ADA – air defense artillery
Atk – attack
Arty – artillery
Avn – aviation
Bde – brigade
EAD – echelons above division
ELINT – electronic intelligence
G2 – assistant chief of staff, intelligence
G3 – assistant chief of staff, operations
HPT – high-payoff target
INFLTREP – in flight report
MLRS – multiple launch rocket system
MI – military intelligence
OP – observation post
P – priority
Recon – reconnaissance
SEAD – suppression of enemy air defense

Figure D-6. Targeting synchronization matrix (sample)

D-16. Both nonlethal and lethal assets may be included in the same matrix. Samples are shown in figures D-7 and D-8. Units may develop their own lethal/nonlethal targeting synchronization matrix.

Example Formats and Target Reports

JFC objective: US/allied nationals, facilities and interests in region protected.
JFC desired effect: Country X unable to affect our ability to generate combat power (4th order effect).
Division desired effect: Enemy X unable to regain control of airfield (33rd order effect).
BCT task: Prior to H-1600 1st BCT disrupts 91st AFF Battalion indirect fire system (81-mm mortar) in area of operations HOG that can place indirect against GERONIMO forward landing site and Route GOLD low water crossing from H-0300 to H+36.
Purpose: To enable 2nd BCT to seize lodgment, build combat power un-impeded, and transition to offensive operations (2nd order effects).
BCT desired effect (end state): No effective enemy fires into the GERONIMO forward landing strip and Route GOLD low water crossings from H-0300 to H+36, when the enemy is expected to infiltrate additional systems from adjacent area of operations.

Decide					Detect		Deliver		Assess			
BCT cdr's desired effects	Targets	CAT	NAI & TAI	Time trigger	Agency	Means	Agency	Means	Agency	Asset	MOP	MOE
Mortar crew/ forward observers unable to function effectively	91st AFF mortar crews and forward obs	FS	TAI 156 157	H-36	BCT intell co	Shadow UAV system defects mortar positions	Spt division/ corps aviation bn	Rotor wing assets support the dissemination of MISOP leaflets to influences crews and forward observers to abandon their positions	BCT intell co	Shadow UAV and HUMINT	Leaflets dropped effectively 30% of mortar positions	91st AFF unable to talk or deliver fires — Thus: 91st AFF (LWC) fires into GERONIMO forward landing site and route GOLD low water crossing disrupted from H-0300 to H+36, when the enemy is expected to infiltrate additional systems from adjacent areas of operations
Local leaders and populace willing and able to supply timely and accurate information	Village leaders and public	Recon	Towns in the vicinity of TAI 156 & 157	H-36 H-24	Bn task force & BCT intell co	Patrols conduct recon and post handbills	Tactical high terminal	Supporting division/ Corps MISOP teams deliver handbills to influence local leaders and populace IPJ110	BCT intell co and MISOP personnel (DS)	HUMINT Patrols	Local leaders and populace report mortar locations and number of deserters that may be mortar crews or forward obs	
Fire support mission command rendered ineffective	91st AFF msn cmd	FS	AO HOG	H-8	BCT intell co	Prophet detects locates, and monitors 91st AAF	BCT intell co and supporting joint forces capabilities	Commando solo net disrupts 91st AFF comm by intrusion on AFF frequency modulation call for fire & mission command net w/msg IJ3210	BCT intell co	Prophet	No comm from 91st AFF detected FS msn cmd effectively disrupted	— Thus: 2nd BCT is able to seize lodgment build combat power unimpeded and transition to offensive operations

Legend:
AO – area of operations
BCT – brigade combat team
bn – battalion
CAT – category
cdr – commander
cmd – command
co – company
comm – communication
DS – direct support
FS – fire support
H – hour (H-hour is the time for a scheduled event to begin.)
HUMINT – human intelligence
intell – intelligence
JFC – joint forces commander
LWC – low water crossing
MISOP – military information support operations
MOE – measure of effectiveness
MOP – measure of performance
msg – message
msn – mission
NAI – named area of interest
Recon – reconnaissance
spt – support
TAI – target area of interest
UAV – unmanned aircraft vehicle
obs – observers
US – United States

Figure D-7. Combined lethal/nonlethal targeting synchronization matrix (sample)

Appendix D

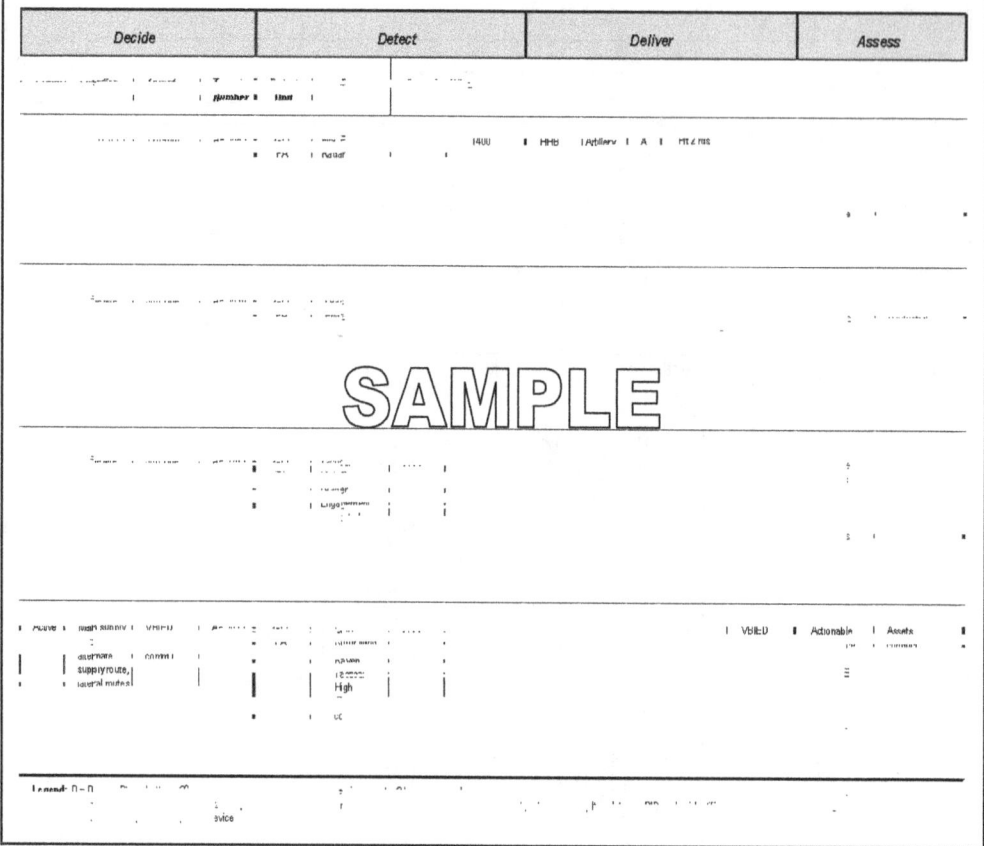

Figure D-8. Alternate targeting synchronization matrix format (sample)

DELIVERY STANDARDS MATRIX

D-17. The sample delivery standards matrix provides criteria for the attack of HPT in each phase of the battle. Units may develop their own delivery standards matrix format. It also facilitates objective decisionmaking for attacking targets at the lowest level possible. Key personnel, such as the field artillery intelligence officer (FAIO), fires cell targeting officer, and the battalion fire direction officer, all refer to the matrix. When HPT are identified they are automatically engaged if they meet the criteria established by the matrix. (See figure D-9.) The matrix provides the following descriptions for each HPT by category for each phase of the operations—

- TLE.
- Size of the target.
- Target activity.
- Time of acquisition.

Example Formats and Target Reports

Category	HPTs	Target Location Error			Size of Unit			Stationary/Moving			Time (Last Verification)		
		Fires Bde, Fires Bn	CAS/ AI	Atk Air	CAS/ AI	Fires Bde, Fires Bn	Atk Air	Fires Bde, Fires Bn	CAS/ AI	Atk Air	Fires Bde, Fires Bn	CAS/ AI	Atk Air
Recon	OPS	100 m – 200 m	200 m	500 m	Sec	Sec	Sec	Stat	Stat	Stat	72 hrs	72 hrs	48 hrs
	Patrols	100 m – 200 m	200 m	1 km	Sec	Sec	Sec	Stat	Stat	Stat/ Move	2 hrs	1 hr	1 hr
	Launch Points	100 m – 200 m	200 m	1 km	Sec	Sec	Sec	Stat	Stat	Stat/ Move	12 hrs	6 hrs	6 hrs
ADA	Hostile Air	100 m – 200 m	200 m		Sec	Sec		Stat	Stat		2 hrs	1 hr	
	Aircraft	100 m – 200 m	200 m		Sec	Sec		Stat	Stat		2 hrs	1 hr	
	Missile Threats	100 m – 200 m	200 m		Sec	Sec		Stat	Stat		2 hrs	1 hr	
FS	Airfields	100 m – 200 m	200 m	500 m	Btry	Bn	Bn	Stat	Stat	Stat	1 hr	2 hrs	2 hrs
	Rockets	100 m – 200 m	200 m	1 km	Btry	Bn	Bn	Stat	Stat	Stat/ Move	1 hr	2 hrs	2 hrs
	Tactical Missiles	100 m – 200 m	200 m	1 km	Btry	Bn	Bn	Stat	Stat	Stat/ Move	1 hr	2 hrs	2 hrs

Legend:
ADA – air defense artillery   Atk Air – attack aircraft   Bde – brigade   Bn – battalion   AI – air interdiction   Btry – battery   CAS – close air support
FS – fire support   HPT – high-payoff target   OP – observation post   Recon – reconnaissance   Sec – section   Stat - stationary

Figure D-9. Delivery standards matrix (sample)

TARGET REPORT

D-18. When targeting information is passed from one agency to another, all essential information must be included to allow for proper analysis and attack. The sample format below will give the targeting working group enough information to properly formulate the best attack response. (See table-D-3.) Units may develop their own target report format.

Appendix D

Table D-3. Target report (sample)

LINE NUMBER
1. Report Agency: COLT
2. Type of Sensor: Human
3. Report DTG: 190044ZNov08
4. Acquisition DTG: 190040ZNov08
5. Distribution: Unknown
6. Posture[1]: Moving
7. Activity[2]: Traveling east towards village
8. Size[3]: 5 white pick trucks
9. Location[4]: 1000010000
10. Location Error[5]: 10 meters

NOTES:
[1] Dug-in, in the open, in built-up areas, and so on.
[2] Moving (direction) or stationary.
[3] Unit size, diameter, and so on.
[4] Grid coordinates
[5] +/- meters

AIR TASKING ORDERS

D-19. The air tasking order is a method used to task and disseminate to components, subordinate units, and other agencies the projected sorties, capabilities, and/or forces to targets and specific missions. Normally provides specific instructions to include call signs, targets, controlling agencies, weapons loads, as well as special instructions. (Joint Publication [JP] 3-30). The joint air tasking cycle is used to develop the air tasking order that articulates the tasking for all joint air operations for a specific execution timeframe, normally 24 hours.

D-20. The air operations center normally establishes a 72- to 96-hour air tasking planning cycle. The battle rhythm or daily operations cycle (schedule of events) articulates suspense's for targeting, air support requests, airspace control means requests, and the air battle plan. The battle rhythm is essential to ensure information is available when and where required to provide products necessary for the synchronization of joint air operations with the joint force commander (JFC) continuity of operations and for supporting other Service components' operations. See table D-4 for an air tasking order example.

Table D-4. Air tasking order (example)

```
EXER/PACIFICA//
MSGID/ATO/505 ECS DOO/ATOAA/FEB///TIMEFRAM/FROM:101200ZFEB2008/TO:111159ZFEB2008//
HEADING/TASKING//TSKCNTRY/DA//SVCTASK/D//
TASKUNIT/1 COMP SQ/ICAO:KHIF//
AMSNDAT/7000/-/-/-/TAL/-/-/DEPLOC:KHIF/ARRLOC:KHIF//
MSNACFT/1/ACTYP:C130H/GOLD08/CP1/-/142/27000/37000//
TASKUNIT/1-3 ATK HEL BN/ICAO:KINS//
AMSNDAT/1601/-/-/-/GATK/-/30M/DEPLOC:KINS/ARRLOC:KINS//
MSNACFT/6/ACTYP:AH64A/BLACKIE01/BEST/-/113/21601/31601//
TASKUNIT/1 FTR SQ/ICAO:KLSV//
AMSNDAT/7245/-/-/-/XCAS/-/-/DEPLOC:KLSV/ARRLOC:KLSV//
TASKUNIT/1 RECCE SQ/ICAO:KLUF//
AMSNDAT/1501/-/-/-/GREC/-/15M/DEPLOC:KLUF/ARRLOC:KLUF//
TASKUNIT/USS BOISE/ICAO:N764//
MTGTLOC/P/8002/-/-/NLT:101202Z/ID:0992E50058DD001/UNK/T/1/-/E19116
/DMPID:373800.0N1181454.0W/-//
TASKUNIT/15TH SPECIAL OPERATIONS SQ/ICAO:KDMA//
AMSNDAT/1300/-/-/-/SOF/-/-/DEPLOC:KDMA/ARRLOC:KDMA// MSNACFT/1/ACTYP:MC130H//
AIRMOVE/1/-/-/-/101530ZFEB/NNN/101531Z//CONTROLA/AWAC/COUSIN/-//
AIRDROP/-/LATS:334959N1180500W/-//
TASKUNIT/319AREFW/ICAO:KSLC//
AMSNDAT/5630/-/-/-/AR/-/-/DEPLOC:KSLC/ARRLOC:KSLC//
MSNACFT/1/ACTYP:KC135R/TEXACO30/TTF/-/111/25630/35630//
AMSNLOC/101520ZFEB/101630ZFEB/CHERRY ANCHOR/250//
REFTSK/BOM/KLBS:077/-/-//5REFUEL/MSNNO/RECCS/NO/ACTYPE/OFLD/ARCT /-//
HEADING/ATO SPECIAL INSTRUCTIONS (SPINS)//
GENTEXT/GENERAL SPINS INFORMATION/
SPINS INDEX
1 GENERAL INFORMATION
1.1. ADDITIONAL INSTRUCTIONS SECTION 15 CONTAINS
1.1.1. TIME SENSITIVE TARGETS
1.1.2. KILL BOX PROCEDURES
1.1.3. INTELLIGENCE, SURVALLIANCE, AND RECONNISANCE
2 COMMANDERS GUIDANCE
2.1. RESTATED MISSION
3 C2 BATTLE MANAGEMENT PLAN
4 RULES OF ENGAGEMENT (ROE)
5 PERSONNEL RECOVERY PROCEDURES
6 COMMUNICATIONS PLAN
7 ELECTRONIC WARFARE PLAN
8 ANTIAIR-WARFARE PLAN NO INPUT
9 THEATER POINT OF CONTACT NO INPUT
10 COMPOSITE AIR OPERATIONS NO INPUT
11 TANKER PROCEDURES
12 STRATEGIC AND THEATER AIRLIFT INSTRUCTIONS NO INPUT
13 SPACE WARFARE
14 AIRLIFT
15 ADDITIONAL INSTRUCTIONS
16 AIRSPACE CONTROL INSTRUCTIONS
```

Appendix D

TARGET INFORMATION FOLDER

D-21. Target information folders have proven to be an efficient and effective way of tracking information related to high-value individuals (HVI). The target information folders normally include a "baseball card," which is a summary of the key information on the HVI. The "baseball card" normally includes the following information—

- Map of HVI area.
- Picture of HVI.
- Personal history of HVI.
- Patterns of life for HVI includes the where, when, who, what, and the how.
- Cell phone number for HVI.
- Car identification.

D-22. The target information folder will also contain additional information as it becomes available—

- Human intelligence reports on the HVI.
- Signal intelligence reports that reference the HVI.
- Imagery/floor plans of likely areas.
- Link diagrams (social/communications networking), both from human intelligence and signal intelligence sources.
- Previous concept of operations targeting the HVI.
- Patrol debriefs.
- Significant activities regarding the HVI.
- Biometrics.

D-23. Figures D-10 through D-15 are samples of the information that may be contained in a target information folder. Units may develop their own format.

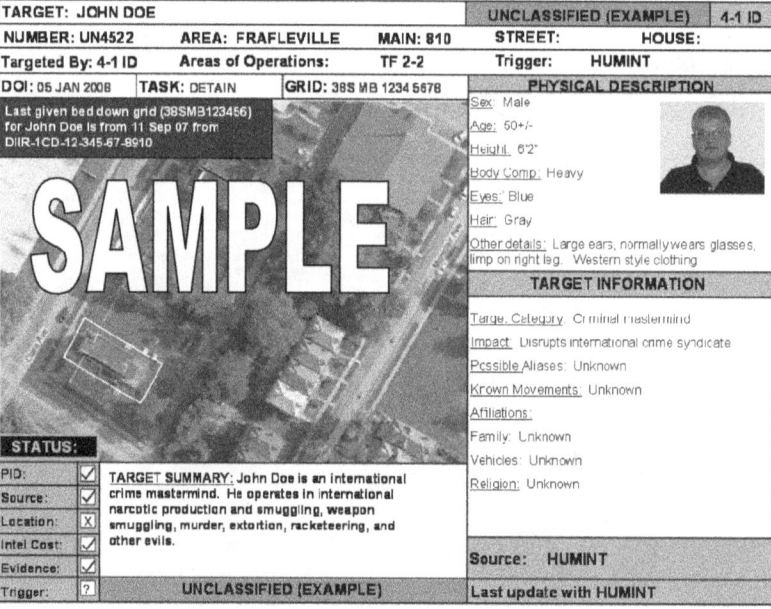

Figure D-10. Baseball card (front side) (sample)

Example Formats and Target Reports

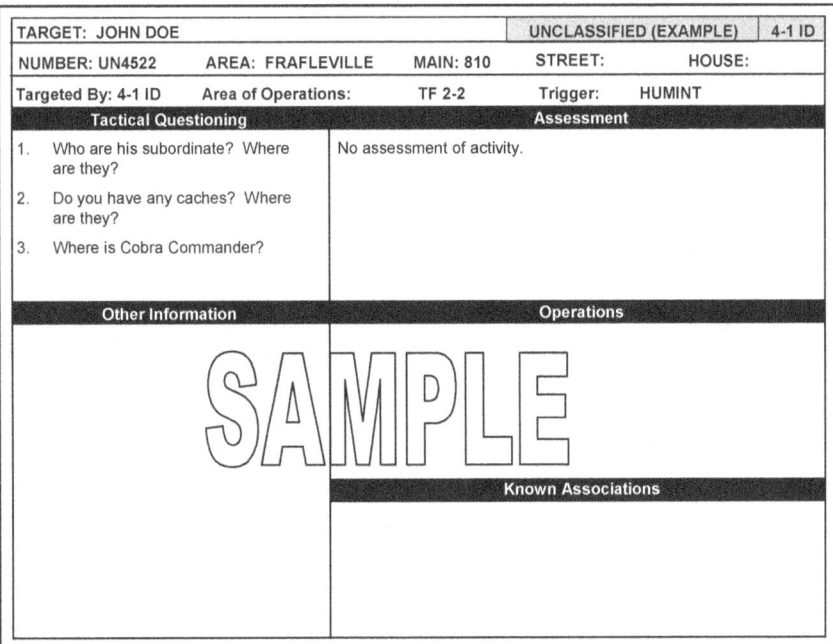

Figure D-11. Baseball card (backside) (sample)

Figure D-12. Picture of HVI residence (sample)

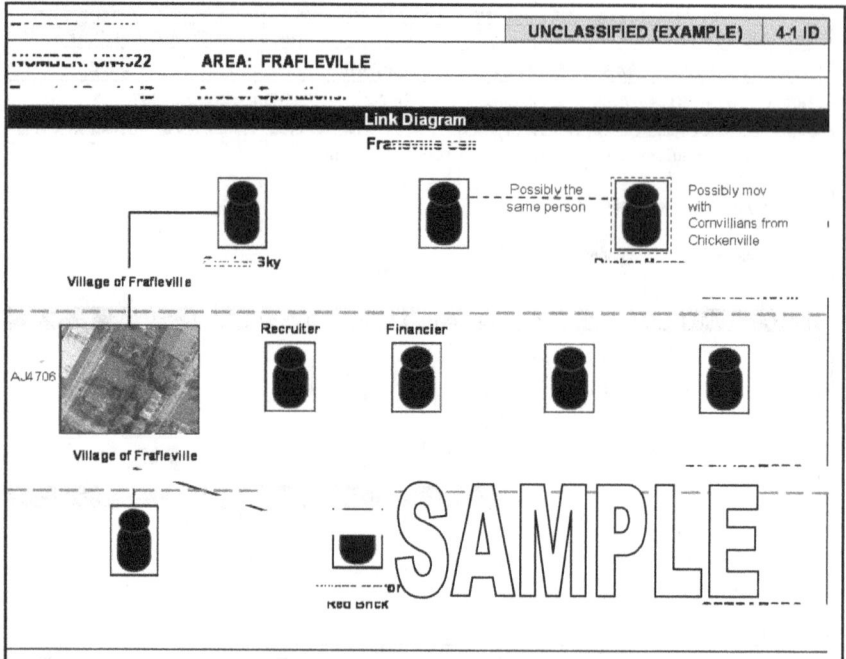

Figure D-13. HVI link diagram (sample)

TARGET: JOHN DOE		UNCLASSIFIED (EXAMPLE)	4-1 ID
NUMBER: UN4522	AREA: FRAFLEVILLE MAIN: 810	STREET:	HOUSE:
Targeted By: 4-1 ID	Area of Operations: TF 2-2	Trigger:	HUMINT
Reporting			
Report Summary	Report Number	Source	DOI
(U/REL EXAMPLE) on 29 Dec 2007 John Doe met with Cobra Commander at Zona Rosa in Kansas City.	SPOT-41D-12-123-12-1234	UNKNOWN	29 Dec 07
(U/REL EXAMPLE) Joe Doe met with Domic Roda on 27 Dec 2007 to discuss IED operations in Lansing.	DIR-4ID-12-123-1234	OTS-123-1234-56	27 Dec 07
	SAMPLE		

Figure D-14. HVI reports (sample)

Example Formats and Target Reports

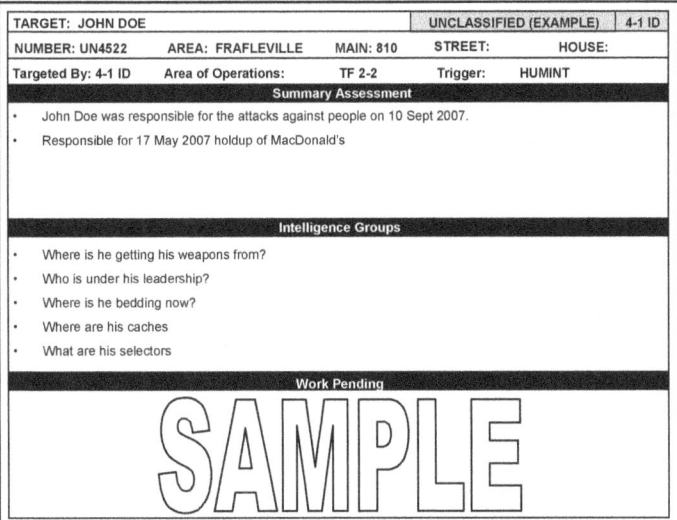

Figure D-15. HVI summary/assessment (sample)

This page intentionally left blank.

Appendix E
Targeting Checklist

DECIDE

_____The commander's planning guidance and intent contain enough detail to enable the targeting working group to determine—
- High-value targets (HVT) to nominate as high-payoff targets (HPT)?
- Desired effects on each HPT?
- When to attack each HPT?
- How to attack each HPT?
- Any restrictions or constraints?
- Which HPT requires battle damage assessment (BDA)?

_____What targeting assets (organic, attached, and supporting) are available to detect and attack HPT?

_____What detect, deliver, and assess support is needed from higher headquarters?

_____When must requests to higher headquarters be submitted to obtain the support required?

_____Have target tracking responsibilities been established?

_____Are systems in place to pass the detected targets to assets that are capable of tracking them?

_____What detect, deliver, and assess support is required from subordinate units, and when is it required?

_____What detect, deliver, and assess support requests have been received from subordinate units, and what has been done with them?

_____Has the AGM been synchronized with the decision support template and the maneuver and fire support plans?

_____Are all commands using a common datum for locations? If not, are procedures in place to correct differences in datum?

DETECT

_____Does the collection plan focus on priority intelligence requirement (PIR) HPT? (This includes HPT designated as PIR.)

_____What accuracy, timeliness, and validity standards target selection standards (TSS) are in effect for detection and delivery systems?

_____Are all target acquisition assets fully employed?

_____Have backup target acquisition systems been identified for HPT?

_____Have responsibilities been assigned to the appropriate unit and/or agency for detection of each HPT?

Appendix E

_____Are HPT being track?

_____Have verification procedures using backup systems been established where necessary?

_____Are target acquisition and BDA requirements distributed properly among systems that can accomplish both?

DELIVER

_____Have communications links been established between detection systems, the decisionmaker, and delivery systems?

_____Have responsibilities been assigned to the appropriate unit and/or agency for attack of each HPT?

_____Has a backup weapon system been identified for each critical HPT? (The primary system may not be available at the time the HPT is verified.)

_____Have fire support coordination measures (FSCM) and/or AGM and clearance procedures been established to facilitate target engagement?

_____Have on order FSCM and/or AGM been established to facilitate future and transition operations?

_____Have potential fratricide situations been identified, and have procedures been established to positively control each situation?

_____Have responsibilities been assigned to the appropriate unit and/or agency for tracking specified HPT and providing BDA on specified HPT?

_____What are the procedures to update the high-payoff target list (HPTL) and synchronize the AGM and decision support template if it becomes necessary to change the scheme of maneuver and fire support as the situation changes?

ASSESS

_____Are the collection assets, linked to specific HPT, still available?

_____Have the collection asset management been notified of the attack of a target requiring assessment?

_____Have the assessment asset managers been updated as to the actual target location?

_____Has all coordination been accomplished for the assessment mission, particularly airborne assets?

_____What is the status of BDA collection?

_____Has the information from the mission been delivered to the appropriate agency for evaluation?

_____Has the targeting working group reviewed the BDA to requests for redirection of air assets?

_____Has the target intelligence gathered from the assessment been incorporated into the overall enemy situational development?

Appendix F
Sample Targeting Working Group Standard Operating Procedures

This appendix provides guidelines for developing a standard/standing operating procedure (SOP) for targeting working groups. Each unit must develop a SOP that is based on the unique mission, organization, equipment, personnel, and philosophy of the commander.

TARGETING WORKING GROUP SOP

F-1. The following example of a targeting working group SOP covers most of the functional areas needing standardization in corps and division targeting working groups. The SOP may be adjusted to serve as a model for brigade and lower echelons.

PURPOSE

F-2. The SOP provides guidelines, routine functions, and to delineate responsibilities for day-to-day operations in the targeting working group.

ORGANIZATION

F-3. The chief of fires leads the targeting working group. In his absence, the assistant chief of staff, operations (G-3) will be the alternate. Membership of the targeting working group routinely consists of representatives from the following staff sections and major subordinate commands. (See figure F-1 below.)

G-2	G-2 Plans/Ops	ALO	DCOF	FAIO
G-3	G-2 Targeting	AVLO	Div/Bde LNO	PSYOP Officer
G-6	G-3 Plans/Ops	COF	DSO	SJA
G-7	ADAM/BAE	CHEMO	Engineer	SOF LNO
G-9	AMD	Collection Manager	EWO	SWO

Legend:
ADAM/BAE – air defense airspace management/brigade aviation element
AMD – air missile defense
ALO – air liaison officer
COF – chief of fires
Div/Bde LNO – division/brigade liaison
EWO - electronic warfare officer
G-2 – assistant chief of staff, intelligence
G-3 – assistant chief of staff, operations
G-6 – assistant chief of staff, signal
G-7 – assistant chief of staff, information engagement
G-9 – assistant chief of staff, civil affairs operations
AVLO – aviation liaison officer
CHEMO – chemical officer
DCOF – deputy chief of fires
DSO – deception staff officer
FAIO – field artillery intelligence officer
LNO - liaison
Ops – operations
PSYOP – psychological operations
SJA – staff judge advocate
SOF LNO – Special operations forces liaison
SWO – staff weather officer

Figure F-1. Targeting working group (example)

CONDUCT

F-4. The targeting working group agenda is divided and briefed during current and future operations. (See figure F-2.) The chief of staff directs the board process and keeps the members focused on the unit mission, commander's intent, targeting guidance, and targeting priorities. The targeting agenda is included in routine staff meetings or drills. Meeting times should be established to allow timely coordination of the parallel targeting effort of senior and subordinate headquarters.

Appendix F

F-5. The staff weather officer begins the session by providing current and predicted weather and its effects on combat operations for the next 72 to 96 hours. Next, the team examines the current situation (present to 24 hours). The G-2 and G-3 brief the enemy and friendly situation with emphasis on current attacks being conducted. The G-2 collection manager briefs battle damage collection currently in effect and possible high-payoff target (HPT) nominations for immediate reattack.

F-6. The current operations agenda involves enemy and friendly situation updates from the G-2 and G-3 that impact on the high-payoff target list (HPTL) and the attack guidance matrix (AGM). They confirm the joint air missions for the following day nominated targets or changes to targets. A significant change in the situation would warrant redirecting allocated joint air capabilities. If there are no significant changes, planning continues for the use of aircraft to support ground operations. The G-3 plans officer briefs a review of operations planning for the next day. G-3 operations brief the concept of operations against the targets assigned. The G-2 collection manager briefs the collection plan to validate targets and pursue battle damage assessment (BDA) based on the target guidance and target priorities. The executors of planned operations brief their respective execution matrixes and conduct any remaining staff coordination needed. The fires cell coordinates the use of Army indirect fires, joint fires, and electronic attacks through the targeting process. This includes the following—

- Coordinate all suppression of enemy air defense (SEAD) missions.
- Develop and recommend fire support coordination measures (FSCM) to support the concept of operations.
- Monitor the status of friendly artillery units.
- Coordinating all Army Tactical Missile System missions controlled by the headquarters.
- Posting the status of friendly maneuver brigades.

F-7. The future operations agenda involves the G-2 and G-3 briefing the anticipated enemy and friendly situations. A review of the war gaming session for this time period is discussed with the chief of staff. Recommended target guidance, target priorities, and objectives are provided to the commander for approval. Targets nominated to support corps and division objectives and priorities are approved and forwarded through channels to the battlefield coordination detachment. G-3 plans briefs shaping operations and attacks for corps or division assets.

What	Who	Why
Current SITREP / CCIR	G-3 Operations	Situation Update
Current Enemy Situation	ASPS	Provide Planning Baseline
Special Staff Considerations	Special Situation	As Requested
Air Status Army Aviation Status	Air Liaison Officer AVN LNO	Update Allocations Review Status and Mission
Collection Plan	AMD	Projected 72-hour Focus
IEW Status	IEW Representative	Review Baseline Priorities
24, 48, 72, and 96- Hour Forecast	G-3 Plans ASPS Targeting Officer	Projected Division Operations Projected Enemy Sets Target Nominations
Approve Nominations: Long-range Target Focus	Chief of Staff	Decision
Review Attack Guidance / HPT	Deputy, Chief of Fires	Validate; Recommend Changes
Final Guidance	Chief of Staff	

Legend:
AMD – air and missile defense
ASPS – all-source production section
ATO – air tasking order
AVN LNO – aviation liaison
CCIR – commander's critical information requirements
G-3 – assistant chief of staff, operations
HPT – high-payoff target
IEW – intelligence electronic warfare
SITREP – situation report
Note: 24, 48, 72, and 96-hours correspond to the ATO cycle. The 24-hour forecast is the current ATO.

Figure F-2. Targeting working group agenda (example)

F-8. Scheduled meetings between corps and division provide an interactive process for the planning and coordinating the allocation of available joint air capabilities during the execution of the joint air tasking cycle (see figure F-3). These meetings synchronize the corps and division current/future operations with the usually at least five joint air tasking orders at any given time for future actions, today's plan, tomorrow's plan, and the day after tomorrow's plan to include the plan in strategy development. The continuous assessment conducted during these meetings allows for the targeting working group to focus on lessons learned, deliberate targeting, and time-sensitive targets (TST) in the detect and deliver function during the D3A methodology.

Sample Targeting Working Group Standard Operating Procedures

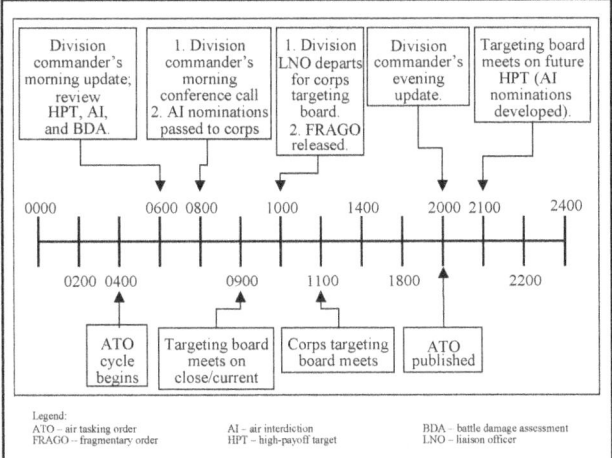

Figure F-3. Meeting times interface between corps and division (example)

RESPONSIBILITIES: CURRENT OPERATIONS AGENDA

F-9. The following paragraphs discuss the responsibilities of individual staff officers and sections for the current operations agenda.

Staff Weather Officer

F-10. The staff weather officer briefs the weather forecast to include light weather data for the next 96 hours. The light weather data impacts friendly air operations for fixed-wing and rotary-wing aircrafts.

Operations

F-11. Briefs recently completed or ongoing attacks and their effects on the current battle. Current operations briefs the following—
- Changes to task organization.
- Current FSCM.
- Relative combat power of all friendly maneuver brigade size units.
- Friendly maneuver unit locations.
- Friendly operations (last 24 hours).
- Friendly scheme of maneuver (next 24 hours).

F-12. Representatives of the executing agencies of each shaping operations (corps and/or division aviation, special operations forces, and others as needed) brief their execution matrix. They may request input or additional guidance from the chief of staff.

Targeting Officer

F-13. The G-2 targeting officer presents BDA obtained from division, corps, and echelons above corps intelligence gathering assets on critical HPT attacked by division, corps, supporting, and subordinate units. Specificity and timeliness are essential. The targeting officer must be proactive in receiving BDA because the degradation of specific targets may be a trigger mechanism for future operations. The corps G-2 and aviation combat element help the targeting officer review and formulate BDA. Target nominations are for immediate reattack of HPT is made at this time.

Appendix F

Plans

F-14. The G-2 plans officer briefs the disposition of important enemy units and associated systems impact on the unit mission. The G-2 also briefs the most likely enemy course of action (COA) in the next 24 to 48 hours.

F-15. The G-3 plans officer reviews the plans for the next 24 to 48 hours that have been handed off to subordinate units for execution. The G-3 briefs the current concept of operations.

Fires Cell

F-16. The deputy chief of fires discusses the approved targeting guidance, HPTL, AGM, and target selection standards (TSS) with the team in light of the G-2 and G-3 situation briefings. The team determines if changes are required.

Collection Management Officer

F-17. The collection management officer reviews the intelligence collection plan for all division, corps, and echelons above corps systems that will assist in targeting. The collection management officer highlights those HPT that cannot be covered with available assets. The guidance is received from the chief of staff on specific or additional requirements.

Air Liaison Officer

F-18. The air liaison officer works closely with the land operations planners to estimate the most likely outcome resulting from employing joint air assets to achieve a specific effect. The following is presented for each target—
- Description.
- Location.
- Type and amount of aircraft to attack (package).
- Ordnance.
- Time on target.
- Any significant changes based on the friendly and/or enemy situation can direct a force packages to operate in a different part of the operational area. It must be approved by the chief of staff because of their knowledge of the complete operational picture.

Information Engagement Officer

F-19. The information engagement officer briefs the inform and influence activities for following—
- Command operational picture.
- Requirements for combat camera, operations security, and civil affairs operations.
- Running estimate for information operations.
- Deconfliction methods for internal and external actions.
- Coordination with outside agencies, higher headquarters, and augmenting forces.

Civil Affairs Officer

F-20. The civil affairs officer briefs the current civilian situation which covers the effect of civilian populations on operations, plans for civilian interference in the area of operations, and the civil affairs mission. Advises the commander on the employment of military units and assets that can support civil affairs operations.

RESPONSIBILITIES: FUTURE OPERATIONS AGENDA

F-21. The following paragraphs discuss the responsibilities of individual staff officers and sections for the future operations agenda. There are three primary briefers for the future operations agenda: G-2 and G-3 plans officers and fires cell representative (usually the deputy chief of fires or and fire support officer

(FSO). Others who might brief during the future operations agenda (depending on the effect targeting has on their mission areas) include the following—
- Aviation liaison officers.
- Fires cell targeting officer.
- Engineer.
- Information engagement officer.
- Deception officer.
- Electronic warfare (EW) officer.
- Special operations forces liaison (if provided).
- Staff judge advocate representative.
- Air defense airspace management/brigade aviation element.
- Liaison officers.
- Civil affairs representative.
- Chemical, biological, radiological, and nuclear officer.
- Psychological operations representative.
- Signal support officer.

Plans Officer

F-22. The G-2 plans officer briefs the disposition of important enemy units and associated systems impacting on the unit mission. He also briefs the most likely enemy COA. This briefing includes enemy follow on forces anticipated to be committed in the unit sector and other forces that will affect future operations. The briefing includes potential HPT that if nominated for attack, meet the commander's intent and if not attacked will significantly impact on future operation plan (OPLAN).

Plans Officer

F-23. The G-3 plans officer will brief any divisional or corps/division operations planned during this time period. The G-3 plans also briefs any branches or sequels to the current OPLAN.

Fires Cell

F-24. The deputy chief of fires discusses Army indirect fires, joint fires, and electronic attacks and presents a decision briefing on proposed targeting guidance and priorities. This includes—
- Recommended target guidance, objectives, and priorities.
- Recommended HPTL, AGM, and TSS.
- Proposed prioritized target list to be forwarded to higher headquarters for execution and targeting tasking for subordinate units.

Aviation Liaison Officer

F-25. The aviation liaison officer answers any questions that the team may have on the capabilities of Army aviation assets. The liaison officer takes the lead in planning attacks on all viable targets with aviation assets.

Information Engagement Officer

F-26. The information engagement officer answers inform and influence activities questions during the operations process. The position targeting responsibilities include—
- Synchronizing appropriate aspects of inform and influence activities with the fires, maneuver, and other warfighting functions.
- Assessing enemy vulnerabilities, friendly capabilities, and friendly missions.
- Nominating inform/influence activities targets for attack.
- Briefing deception operations.

Appendix F

- Providing operation security measures.
- Synchronizing inform and influence activities.

Targeting Officer

F-27. The targeting officer prepares information briefings for the deputy chief of fires to include the following—
- Targeting guidance and priorities.
- The targeting working group meeting agenda.
- HPTL, AGM, and TSS.

F-28. The HPTL include HPT nominations submitted by the aviation combat element and subordinate units to the unit fires cell. The targets are prioritized based on approved targeting guidance and priorities. The targeting officer also updates the situation map and provides the team with all current and proposed FSCM. The targeting officer is also responsible for consolidating, coordinating, providing to the team for approval and disseminating the restricted, and no-strike target list. The list includes historical, religious, educational, civic, and humanitarian sites within the unit boundaries.

Engineer Officer

F-29. The unit engineer provides expertise on enemy capabilities for bridging, breaching, and infrastructure construction. The position helps the targeting working group determine target feasibility of enemy engineer equipment. Specifically, the engineer representative must be prepared to discuss such things as the following—
- The width of gap the enemy can bridge.
- The depth of any minefields the enemy can breach and location of breach sites.
- The ability of the enemy to repair bridges, roads, airfields, and ports.
- The obstacles plan is included in target planning.

Deception Officer

F-30. The deception officer advises the team on conflicts between targeting and deception plans.

Electronic Warfare Officer

F-31. The electronic warfare (EW) officer advises the working group members on the capability and availability of all EW assets.

Special Operations Forces Liaison

F-32. The special operations forces liaison or liaison element (when provided) advises the board on special operations forces missions in the area of operations and their capabilities as they relate to targeting. The special operations element also helps formulate FSCM established to protect special operations forces.

Staff Judge Advocate Representative

F-33. The staff judge advocate's representative on the targeting working group will provide analysis and advice throughout the planning process to ensure compliance with the rules of engagement and all applicable laws.

Air and Missile Defense Officer

F-34. The air and missile defense officer is responsible for deconflicting airspace management and coordination.

Liaison Officer

F-35. The division and/or brigade liaison officer addresses the concerns of their commanders pertaining to targeting and future operations. They are prepared to discuss their commander's targeting priorities, future plans, and air interdiction target nominations. The discussion prepares the unit staff to support and anticipate the targeting needs of subordinate units. The liaison officer is prepared to discuss updates to FSCM.

Civil Affairs Representative

F-36. The staff representative verifies the protected and restricted target list and helps the board answer questions on collateral damage issues. Civil affairs conduct detailed assessments of the local population and the area of operations. The assessments are used to provide information on which targets might have positive or negative effect on morale or infrastructure and logistics systems of the enemy. The representative also advises on the expected number and direction of flow of dislocated civilians and how they will interfere with military operations.

Chemical Officer

F-37. The unit chemical officer provides expertise on the weapons of mass destruction capability of the enemy. The chemical officer helps the targeting working group determine target feasibility of the weapons. The officer advises on the impact that facilities (employment, storage, and production) would have on the battlefield and friendly operations, if attacked. The chemical position provides guidance on the employment of smoke and obscurants and their impact on weapon systems and sensors.

Psycholoical Operations Officer

F-38. The psychological operations officer analyzes potential targets based on their significance in accomplishing a specific mission. The officer selects targets that are susceptible to military information support operations and participates in the target nomination process to include coordinating available assets to engage the targets.

Signal Support Officer

F-39. The signal officer provides expertise on the employment of friendly information systems to include advice on the integration of the five signal support functions. The functions are as follows—
- Communications.
- Automation.
- Visual Information.
- Printing and Publications.
- Records Management.

F-40. The five functions provide a full functioning, synchronized information system. The signal officer coordinates with the chief of staff, G-3, and other targeting working group members as required.

This page intentionally left blank.

Appendix G
Common Datum

For joint agencies to coordinate targeting functions properly, they must be able to exchange information by using a common frame of reference regarding the operational area. A small detail that has tremendous implications supporting this common reference, especially if overlooked, is ensuring planners and operators use the correct datum.

During the first days of Operation Desert Storm, the Air Force reported that the Boeing B-52 Stratofortress raids consistently fell short of the target. The weapon systems locating targets were on a different datum from the Boeing B-52 navigation system. The B-52 bomber missions were successful after the datum issues were identified and corrected. In some cases, forward observers, fire direction centers, and/or weapon systems were using issued maps with different datums. Location variances exceeded 700 meters. In other Operation Desert Storm examples, a global positioning system receiver position was compared to a paper map position and in many cases the positions varied by hundreds of meters. In another example, troops deploying to Somalia found the maps available were a mix of Russian maps found in Gauss Kruger and World Geodetic System 72.

All joint operations utilize the World Geodetic Systems 84 datum for references coordinates according to the Chairman Joint Chiefs of Staff Instruction 3900.01C.

DATUM AND PROJECTION AWARENESS

G-1. Personnel working in any organization that exchanges information regarding position locations must be aware of the need for using a common datum. In the past, this was not viewed as critical because weapons technology and employment doctrine did not require highly accurate positions. The possibility of deployment to many new foreign locations, where the only maps are on local datum and foreign projections, make precise coordinates vital to mission success. The problem is even more critical with the advent of growing technology (for example, global positioning system, tactical land attack missile system, and so forth). Ignoring the map datum could result in fratricide or gross target location errors.

G-2. A datum is the mathematical model of the earth used to calculate position coordinates on any map, chart, or survey system. Many countries use their own datum(s) when they make their maps and survey. The *local datums* may produce coordinates that vary greatly from datums used by United States forces. Different datums are used even within the United States Department of Defense.

G-3. Presently there are more than 1,000 datums in use. The National Geospatial-Intelligence Agency is concerned with only 200 of these, which are found on paper maps, digital mapping products, and other products provided to the user. As datums are shifted to a common grid, the change in the location of a unit may be more than 1,000 meters. There are different datums for both horizontal and vertical reference. Most vertical datums use mean sea level for elevation, but they may use any of over a hundred different horizontal position datums. The agency is reducing the number of datums used for their products. As map products are updated, they will be updated with standard datums and ellipsoids. Do not mix old and new products. The National Geospatial-Intelligence Agency decided to use world geodetic system 84 in most of the world. During transition, some paper products will not match digital products.

Appendix G

G-4. There can be several error possibilities in air ground operations if multiple datums are used. A few are listed below—
- Friendly position location errors.
- Enemy target location errors.
- Combat search and rescue location errors.
- Navigation aid use.
- Aircraft vectoring errors.
- Airspace control measure errors.
- Air defense errors.

DATUM USE

G-5. The joint force commander (JFC) must identify which datum information is to be used within the joint force for exchange of position information. Subordinate personnel must know the answer to these questions—
- What is the operational datum?
- On what datum are the target coordinates?
- On what datum are the position coordinates?
- On what datum are fire control systems?
- On what datum are the current operational maps?

CHECKING DATUM AND CONVERTING COORDINATES

G-6. Datums can be checked. All maps and products that the National Geospatial-Intelligence Agency distributes have the datum printed somewhere in the margin. The figure below is an example of typical datum information provided in the margin of a map. Any element who converts coordinates from latitude and longitude to Universal Transverse Mercator or military grid reference system, and the reverse requires National Geospatial-Intelligence Agency-Mapping Datum Transformation software to perform conversions. It can also transform coordinates between world geodetic system 84 and over a hundred other datums. The mapping datum transformation software comes with instructions. The software works on any International Business Machines personal computer compatible computer with an external drive. Purchase the software from wherever the user procures National Geospatial-Intelligence Agency products by using stock number Mapping Datum Transformation International Business Machines Personal Computer.

```
ELLIPSOID..................................................WORLD GEODETIC SYSTEM 1984
GRID.................................................................1,000 METERS UTM ZONE 51
PROJECTION.........................................................TRANSVERSE MERCATOR
VERTICLE DATUM..........................................................MEAN SEA LEVEL
HORIZONTAL DATUM...............................................WORLD GEOTIC SYSTEM 1984
HYDROGRAPHIC DATUM...............................APPROXIMATE LEVEL OF LOWEST
                                LOW WATER
PRINTED BY.............................................................................DMAHTC 4-94

                    COODINATE CONVERSIOT WGS 84 TO TOKYO
                       GRID: Add 156m.E; Subtract 712m.N.
                    Geographic: Add 7.0"Long.; Subtract 8.5"Lat.
```

Figure G-1. Map margin datum (example)

Appendix H
Target Numbering

H-1. The target numbering system is a writing system for expressing lethal and nonlethal actions against an entity or object considered for possible engagement or other actions. This numbering system identifies a wide array of mobile and stationary forces, equipment, capabilities, and functions that an enemy commander can use to conduct operations. The Army utilizes alphanumeric characters for the targeting process when selecting and prioritizing targets, and matching the appropriate response, and considering operational requirements and capabilities. The Army assigns target numbers that adheres to the provisions of STANAG 2934.

H-2. The targeting alphanumeric system represents the following—
- Organizations.
- Elements.
- Cell, sections, or teams within a brigade size element.
- Cell, sections, or teams within a battalion size element.
- Block of numbers.

H-3. The target number is comprised of six characters consists of two letters and four numbers in the following positions, for example AB1234. The two letters indicate the originator of the target number and/or the echelon holding the target data. The letter "Z" is the only permanently assigned first letters. The senior headquarters establishes and publishes in the operations order the assigned first letter. The target number prefix "Z" is reserved for technical use by automatic data processing systems among nations when transferring target information from one nation to another. The automation system will use a target number beginning with the prefix "Z." The second letter "E" is allocated for Service components forces in automatic data processing systems in those instances where a "Z" prefix target is generated for example "ZE." Table H-1 is an example of the assignment of first letters for targeting in an operational environment.

Table H-1. Assignment of first letter (example)

Organization	Letter
CENTCOM	C
United States	A
United Kingdom	B
EUROCORPS	E
France	F
Germany	D
MNC	M
NRDC Italy	N

Legend:
CENTCOM — Central Command
EUROCORPS — European Corps
MNC — Multi National Corps
NRDC — NATO Rapid Deployable Corps

H-4. Target numbers serve as an index to all other information regarding a particular target, such as location, description, and size. Normally, a common target numbering system is used at corps and within a major force. Target block numbers are traceable to its originating source to specific users. Corps down to brigade level may assign the second letter (A through Z). See table H-2.

Appendix H

Table H-2. Assignment of letters (example)

Elements	Letters
Corps	AA
Fires Cell	AB
TACP	AC
X Division	AD
1 BCT	AE
2 BCT	AF
3 BCT	AG
4 BCT	AH
Y Division	AJ
1 BCT	AK
2 BCT	AM
3 BCT	AN
4 BCT	AQ

Legend: TACP – tactical air control party
BCT – brigade combat team

H-5. See Table H-3 for an example of standard blocks of numbers assigned within a brigade.

Table H-3. Assignment of blocks of numbers (example)

Numbers	Brigade elements
0000-2999	BCT Fires Cell and COLT
3000-3999	Fires cell, lowest numbered maneuver battalion or squadron[1]
4000-4999	Fires cell, second lowest numbered maneuver battalion or squadron
5000-5999	Fires cell, third lowest numbered maneuver battalion or squadron
6000-6999	Additional Fires cells or fire support assets
7000-7999	FDC, BCT fires battalion
8000-8999	Counterfire targets
9000-9999	Spare

Legend: [1] Lowest regimental number   BCT - brigade combat team
COLT – combat observation and lasing team   FDC – fire direction center

H-6. The battalion size element with a block of numbers may allocate numbers as shown in table H-4. Consult the unit standing operating procedure (SOP) for specific unit target numbers. Additional number blocks are requested from the supervising fires cell.

Table H-4. Additional assignment of blocks of numbers (example)

Numbers	Battalion elements
X000-X199	Battalion Fires Cell
X200-X299	FIST, Company A
X300-X399	FIST, Company B
X400-X499	FIST, Company C
X500-X599	FIST, Company D
X600-X699	Additional FIST or fire support assets
X700-X799	FDC, battalion or company mortars
X800-X999	Spare

Legend: FDC – fire direction center FIST – fire support team
X – numeral assigned by higher HQ

Glossary

The glossary lists acronyms and terms with Army, multi-Service, or joint definitions, and other selected terms. Where Army and joint definitions are different, "(Army)" follows the term. Terms for which Field Manual (FM) 3-60 is the proponent manual (the authority), are marked with an asterisk (*). The proponent manual for other terms is listed in parentheses after the definition. Terms for which the Army and Marine Corps have agreed on a common definition are followed by "(Army-Marine Corps)"

SECTION I – ACRONYMS AND ABBREVIATIONS

AGM	attack guidance matrix
BCT	brigade combat team
BDA	battle damage assessment
COA	course of action
COLT	combat observation and lasing team
CP	command post
D3A	decide, detect, deliver, and assess
DJIOC	Defense Joint Intelligence Operations Center
EW	electronic warfare
F2T2EA	find, fix, track, target, engage, and assess
F3EAD	find, fix, finish, exploit, analyze, and disseminate
FAIO	field artillery intelligence officer
FIB	fires brigade
FM	field manual
FMI	field manual interim
FSCM	fire support coordination measure
FSO	fire support officer
G-2	assistant chief of staff, intelligence
G-3	assistant chief of staff, operations
HPT	high-payoff target
HPTL	high-payoff target list
HVI	high-value individual
HVT	high-value target
IPB	intelligence preparation of the battlefield
JIPOE	joint intelligence preparation of the operational environment
ISR	intelligence, surveillance, and reconnaissance
J-2	intelligence directorate of a joint staff
J-3	operations directorate of a joint staff
JFACC	joint force air component commander
JFC	joint force commander
JFLCC	joint force land component commander

JP	joint publication
J-SEAD	joint suppression of enemy air defense
JTCB	joint targeting coordination board
JTF	joint task force
MDMP	military decisionmaking process
METT-TC	mission, enemy, terrain and weather, troops and support available, time available, civil considerations
MOE	measure of effectiveness
MOP	measure of performance
OPLAN	operation plan
OPORD	operation order
PIR	priority intelligence requirement
S-2	intelligence staff officer
S-3	operations staff officer
S-7	information engagement staff officer
S-9	civil affairs staff officer
SEAD	suppression of enemy air defense
SOP	standard/standing operating procedure
TACP	tactical air control party
TLE	target location error
TSS	target selection standard
TST	time-sensitive target

SECTION II – TERMS

air interdiction

(joint) Air operations conducted to divert, disrupt, delay, or destroy the enemy's military potential before it can be brought to bear effectively against friendly forces, or to otherwise achieve objectives. Air interdiction is conducted at such distance from friendly forces that detailed integration of each air mission with the fire and movement of friendly forces is not required. (JP 1-02)

airspace control authority

(Army) The commander designated to assume overall responsibility for the operation of the airspace control system in the airspace control area. Also called **ACA**. (FM 3-52.2)

Army air-ground system

(Army) The Army system which provides for interface between Army and tactical air support agencies of other Services in the planning, evaluating, processing, and coordinating of air support requirements and operations. It is composed of appropriate staff members, including G-2 air and G-3 air personnel, and necessary communication equipment. (FM 3-52.2)

assessment

(joint) 1. A continuous process that measures the overall effectiveness of employing joint force capabilities during military operations. 2. Determination of the progress toward accomplishing a task, creating an effect, or achieving an objective. 3. Analysis of the security, effectiveness, and potential of an existing or planned intelligence activity. 4. Judgment of the motives, qualifications, and characteristics of present or prospective employees or "agents." (JP 1-02)

Glossary

*attack guidance matrix

    (Army) A matrix, approved by the commander, which addresses which targets will be attacked, how, when, and the desired effects.

civil affairs

    (joint) Designated Active and Reserve component forces and units organized, trained, and equipped specifically to conduct civil affairs operations and to support civil-military operations. Also called **CA**. (JP 1-02)

combat assessment

    (joint) The determination of the overall effectiveness of force employment during military operations. Combat assessment is composed of three major components: (a) battle damage assessment; (b) munitions effects assessment; and (c) reattack recommendation. (JP 1-02)

*desired effects

    (Army) The damage or casualties to the enemy or material that a commander desires to achieve from an identical target engagement. Damage effects on material are classified as light, moderate, or severe.

electronic attack

    (joint) Division of electronic warfare involving the use of electromagnetic energy, directed energy, or antiradiation weapons to attack personnel, facilities, or equipment with the intent of degrading, neutralizing, or destroying enemy combat capability and is considered a form of fires. Also called **EA**. (JP 1-02)

electronic warfare

    (joint) Military action involving the use of electromagnetic and directed energy to control the electromagnetic spectrum or to attack the enemy. Electronic warfare consists of three divisions: electronic attack, electronic protection, and electronic warfare support. Also called **EW**. (JP 1-02)

fires

    (joint) The use of weapon systems to create a specific lethal or nonlethal effect on a target. (JP 1-02)

high-payoff target

    (joint) A target whose loss to the enemy will significantly contribute to the success of the friendly course of action. High-payoff targets are those high-value targets that must be acquired and successfully attacked for the success of the friendly commander's mission. Also called **HPT**. (JP 1-02)

*high-value individual

    A high-value individual is a person of interest (friendly, adversary, or enemy) who must be identified, surveilled, tracked and influenced through the use of information or fires. An HVI may become a high-payoff target (HPT) that must be acquired and successfully attacked (exploited, captured, or killed) for the success of the friendly commander's mission.

high-value target

    (joint) A target the enemy commander requires for the successful completion of the mission. The loss of high-value targets would be expected to seriously degrade important enemy functions throughout the friendly commander's area of interest. Also called **HVT**. (JP 1-02)

intelligence, surveillance, and reconnaissance

    (joint) An activity that synchronizes and integrates the planning and operation of sensors, assets, and processing, exploitation, and dissemination systems in direct support of current and future operations. This is an integrated intelligence and operations function. (JP 1-02) [Note: the Army definition adds: "For Army forces, this activity is a combined arms operation that focuses primarily on priority intelligence requirements while answering the commander's critical information requirements."] Also called **ISR**. (FM 1-02/MCRP 5-12A)

interdiction

    (joint). An action to divert, disrupt, delay, or destroy the enemy's military surface capability before it can be used effectively against friendly forces, or to otherwise achieve objectives. (JP 1-02)

Glossary

joint targeting coordination board

(joint) A group formed by the joint force commander to accomplish broad targeting oversight functions that may include but are not limited to coordinating targeting information, providing targeting guidance and priorities, and refining the joint integrated prioritized target list. The board is normally comprised of representatives from the joint force staff, all components, and if required, component subordinate units. Also called **JTCB**. (JP 1-02)

measure of effectiveness

(joint) A criterion used to assess changes in system behavior, capability, or operational environment that is tied to measuring the attainment of an end state, achievement of an objective, or creation of an effect. Also called **MOE**. (JP 1-02)

measure of performance

(joint) A criterion to assess friendly actions that is tied to measuring task accomplishment. Also called **MOP**. (JP 1-02)

no-strike list

(joint) A list of objects or entities characterized as protected from the effects of military operations under international law and/or rules of engagement. Attacking these may violate the law of armed conflict or interfere with friendly relations with indigenous personnel or governments. Also called **NSL**. (JP 1-02)

priority intelligence requirement

(joint) An intelligence requirement, stated as a priority for intelligence support, that the commander and staff need to understand the adversary or operational environment. (JP 1-02) (Marine Corps) An intelligence requirement associated with a decision that will critically affect the overall success of the command's mission. Also called **PIR**. (MCRP 5-12C)

restricted target list

(joint) A list of restricted targets nominated by elements of the joint force and approved by the joint force commander. This list also includes restricted targets directed by higher authorities. Also called **RTL**. (JP 1-02)

tactical air control party

(Army/Marine Corps) A subordinate operational component of a tactical air control system designed to provide air liaison to land forces and for the control of aircraft. (Marine Corps) A subordinate operational component of a tactical air control system organic to infantry divisions, regiments, and battalions. Tactical air control parties establish and maintain facilities for liaison and communications between parent units and airspace control agencies, inform and advise the ground unit commander on the employment of supporting aircraft, and request and control air support. Also called **TACP**. (FM 1-02/MCRP 5-12A)

target

(joint) 1. An entity or object considered for possible engagement or other action. (JP 1-02, part 1 of a 4-part definition)

target area of interest

(joint) The geographical area where high-value targets can be acquired and engaged by friendly forces. Not all target areas of interest will form part of the friendly course of action; only target areas of interest associated with high priority targets are of interest to the staff. These are identified during staff planning and wargaming. Target areas of interest differ from engagement areas in degree. Engagement areas plan for the use of all available weapons; target areas of interest might be engaged by a single weapon. Also called **TAI**. (JP 1-02)

target development

(joint) The systematic examination of potential target systems — and their components, individual targets, and even elements of targets — to determine the necessary type and duration of the action that must be exerted on each target to create an effect that is consistent with the commander's specific objectives. (JP 1-02)

*target selection standards
    Target Selection Standards (TSS) are criteria, applied to enemy activity (acquisitions and battlefield information), used in deciding whether the activity is a target.

targeting
    (joint) The process of selecting and prioritizing targets and matching the appropriate response to them, considering operational requirements and capabilities. (JP 1-02)

validation
    (joint) 2. A part of target development that ensures all vetted targets meet the objectives and criteria outlined in the commander's guidance and ensures compliance with the law of armed conflict and rules of engagement. (JP 1-02, part 2 of a 4-part definition)

vetting
    (joint) A part of target development that assesses the accuracy of the supporting intelligence to targeting. (JP 1-02)

weaponeering
    (joint) The process of determining the quanity of a specific type of lethal or nonlethal weapons required to achieved a specific level of damage to a given target considerating target vulnerability, weapons characteristics and effects, and delivery parameters. (JP3-60)

This page intentionally left blank.

References

Reference military publications are listed by title. Most joint publications are available online: <http://www.dtic.mil/>. Most Army doctrinal publications are available online: <http://www.army.mil/usapa/index.html>.

REQUIRED PUBLICATIONS

FM 1-04, *Legal Support to the Operational Army*, 15 April 2009.

FM 3-05.40, *Civil Affairs Operations*, 29 September 2006.

FM 3-13, *Information Operations: Doctrine, Tactics, Techniques, and Procedures*, 28 November 2003.

FM 3-36, *Electronic Warfare in Operations*, 25 February 2009.

JP 2-0, *Joint Intelligence*, 22 June 2007.

JP 3-03, *Joint Interdiction*, 3 May 2007.

JP 3-13.1, *Electronic Warfare*, 25 January 2007.

JP 3-60, *Joint Targeting*, 13 April 2007.

TC 2-01, *Intelligence, Surveillance, and Reconnaissance (ISR) Synchronization*, 22 September 2010.

RELATED PUBLICATIONS

JOINT AND DEPARTMENT OF DEFENSE PUBLICATIONS

JP 1-02, *Department of Defense Dictionary of Military and Associated Terms*, 12 April 2001.

JP 2-01, *Joint and National Intelligence Support to Military Operations*, 7 October 2004.

JP 2-01.3, *Joint Intelligence Preparation of the Operational Environment*, 16 June 2009.

JP 2-03, *Geospatial Intelligence Support to Joint Operations*, 22 March 2007.

JP 3-0, *Joint Operations*, 17 September 2006.

JP 3-13, *Information Operations*, 13 February 2006.

JP 3-14, *Space Operations*, 6 January 2009.

JP 3-30, *Command and Control for Joint Air Operations*, 12 January 2010.

ARMY PUBLICATIONS

ATTP 3-09.13, *Battlefield Coordination Detachment*, 21 July 2010.

ATTP 3-09.36, *Army Fires Observer*, 17 February 2010.

FM 1-02, *Operational Terms and Graphics*, 21 September 2004.

FM 3-0, *Operations*, 27 February 2008.

FM 3-05.401, *Civil Affairs Tactics, Techniques, and Procedures*, 5 July 2007.

FM 3-24, *Counterinsurgency*, 15 December 2006.

FM 3-52.2, *Multi-Service Tactics, Techniques, and Procedures for the Theater Air Ground System*, 10 April 2007.

FM 3-61.1, *Public Affairs Tactics, Techniques, and Procedures*, 01 October 2000.

FM 5-0, *The Operations Process*, 26 March 2010.

FM 5-19, *Composite Risk Management*, 21 August 2006.

FM 6-20-40, *Tactics, Techniques, and Procedures for Fire Support for Brigade Operations (Heavy)*, 5 January 1990.

FMI 2-01.301, *Specific Tactics, Techniques, and Procedures and Applications for Intelligence Preparation of the Battlefield*, 31 March 2009.

MARINE CORPS PUBLICATIONS

MCRP 5-12C, *Marine Corps Supplement to the Department of Defense Dictionary of military and Associated Terms*, 23 July 1998.

MCWP 3-16, *Fire Support in the Ground Combat Element*, 28 November 2001.

NATO PUBLICATIONS

STANAG 2934 (Ed. 3), *Artillery Procedures*, 27 April 2009.

PRESCRIBED FORMS

None

REFERENCED FORMS

DA Form 2028 Recommended Changes to Publications and Blank Forms.

Index

A
AGM, 1-9, 2-2, 2-5, 2-9, 2-20, 3-6, D-3
air tasking order, D-12
assess, 1-10, 2-19, A-4, D-7, E-2
assessment, 2-19
attack guidance, 2-8, 2-15, 4-1, 4-6
attack guidance matrix, 1-9, 2-2, 2-5, 2-9, 2-20, 3-6, D-3

B
battalion/squadron
  assistant FSO, 4-22
  fires cell, 4-21
  FSO, 4-21
battle damage assessment, 2-20, 3-6
BCT
  company/troop
    fire support teams, 4-21
BCT battalion
  joint terminal attack controller, 4-23
  terminal attack control for close air support, 4-23
BCT battalion/squadron
  fire support sergeant, 4-23
  fire support specialist, 4-23
  fire support/targeting noncommissioned officer, 4-22
  fires cell, 4-21
BCT brigade judge advocate, 4-20
BCT command and staff
  S-7, 4-17
  S-9, 4-18
BCT fires cell
  information coordinator, 4-17
BCT fires cell
  brigade operational law team, 4-20
  civil affairs officer, 4-18
  fire support noncommissioned officer, 4-14
  fire support specialist, 4-15
  Marine Corps liaison officer, 4-17
  military information support operations noncommissioned officer, 4-18
  naval surface fire support liaison officer, 4-16
  public affairs broadcast specialist, 4-19
  public affairs noncommissioned officer, 4-19
  public affairs specialist, 4-19
  tactical air control party, 4-16
  target analyst/targeting noncommissioned officer, 4-15
BCT fires cell operations noncommissioned officer, 4-14
BCT fires cell public affairs officer, 4-19
BDA, 2-20, 3-6
brigade fire support officer, 4-25
brigade fire support officer, 4-8, 4-11, 4-24
brigade FSO, 4-8, 4-11, 4-24, 4-25

C
combat observation and lasing team, 4-15
corps and division targeting, 3-1

D
datum, G-1
deceive, 1-3
decide, 1-8, 1-9, 2-2, B-1
degrade, 1-3
delay, 1-3
deliberate targeting, 1-5, A-1
deliver, A-1, B-2
deny, 1-3
destroy, 1-3
detect, 1-9, 2-10, A-1, B-2
disrupt, 1-3
divert, 1-4
dynamic targeting, 1-5, A-1

E
effects
  deceive, 1-3
  degrade, 1-3
  delay, 1-3
  deny, 1-3
  destroy, 1-3
  disrupt, 1-3
  divert, 1-4
  exploit, 1-4
  interdict, 1-4
  neutralize, 1-4
  suppress, 1-4
electronic warfare officer, 3-14, 4-17, F-6
exploit, 1-4

F
FAIO, 2-6, 3-6, 3-13
field, 4-5
field artillery intelligence officer, 3-13
field artillery intelligence officer, 2-6, 3-6
fire support coordinator, 4-5, 4-8
fire support officer, 4-21
fires cell, 4-21
fires cell, 3-1, 3-16, 4-11

H
high-payoff target, 2-2, 2-4, 2-5
high-payoff target list, 1-9, 2-2, 2-5, 4-1, 4-8, 4-27, D-1, D-3
high-payoff target list, 3-6
high-payoff target list, D-3
high-value individual, B-1, D-14
high-value target, 2-2, 2-3, 2-4
HPT, 2-2, 2-4, 2-5
HVI, B-1, D-14
HVT, 2-2, 2-3, 2-4

I
information operations, 1-3, C-3
intelligence preparation of the battlefield, 2-3, 2-7
interdict, 1-4
IPB, 2-3, 2-7

Index

J
joint targeting cycle, 3-1, 3-5
joint targeting process, 3-1, 3-5

N
named area of interest, 2-4
neutralize, 1-4

O
on-call target, 1-5

P
physical damage assessment, 2-21
planned targets, 1-5

S
scheduled target, 1-5
sensitive target, 1-6
suppress, 1-4

T
tactical air control party, 3-18
target area of interest, 2-4
target development, 2-13
target selection standard, D-2, D-6
target selection standards, 2-7
target validation, 2-14
target value analysis, 2-4
target vetting, 2-14
targeting, 3-11
targeting board, 4-4
targeting officer, F-6
targeting officer, 3-13, 4-9
targeting officer, 2-11
targeting officer, 4-13
targeting working group, 1-11
targeting working group, 2-6
targeting working group, 2-10
targeting working group, 2-12
targeting working group, 2-16
targeting working group, 2-18
targeting working group, 2-20
targeting working group, 2-21
targeting working group, 4-2
targeting working group, 4-3
targeting working group, 4-4
targeting working group, 4-24
targeting working group conducting sessions of, 4-25
targeting working group information provided by core members, 4-25
targeting working group subsequent actions, 4-27
targeting working group, E-1
targeting working group, F-1
theater air-ground systems, 3-9
time-sensitive target, 1-5, A-1
TSS, 2-7, D-2, D-6
TST, 1-5, A-1

U
unanticipated target, 1-5
unplanned target, 1-5

V
validation, 2-14
vetting, 2-14

W
weaponeering, 2-17

www.ingramcontent.com/pod-product-compliance
Lightning Source LLC
Chambersburg PA
CBHW081323040426
42453CB00013B/2288